T0214028

Principles of Electrical Neural Interfacing

Liang Guo

Principles of Electrical Neural Interfacing

A Quantitative Approach to Cellular Recording and Stimulation

 Springer

Liang Guo
The Ohio State University
Columbus, OH, USA

ISBN 978-3-030-77679-4   ISBN 978-3-030-77677-0   (eBook)
https://doi.org/10.1007/978-3-030-77677-0

This Springer imprint is published by the registered company Springer Nature Switzerland AG
The registered company address is: Gewerbestrasse 11, 6330 Cham, Switzerland

To my lovely daughters, Stella and Gillian

Preface

Electrical neural recording and stimulating technologies are the two foundational interfacing techniques in the electrical neuroengineering area, upon which numerous applications in neuroelectrophysiology and neuroprosthetics are built. Electrical neural interfacing employs solid-state electrodes to record or stimulate on neural tissues, such as the brain, spinal cord, nerves, and muscles. Its invention, development, and application closely accompanied the emergence, growth, and prosperity of the field of neuroengineering. It has also been a powerful engine to propel the broader field of electrophysiology for about a century. Rooting in the disciplines of electrical engineering and applied physics with a focus on nervous systems, it was, and sometimes still is, an amusing branch of these disciplines. Only in more recent decades has it been recategorized under the new field of neuroengineering by a growing number of institutions worldwide. However, as I haven't yet witnessed the establishment of a neuroengineering department in any school so far, presently neuroengineering either exists as an interdisciplinary program across multiple traditional disciplines or is organized in the department of electrical engineering, biomedical engineering, or bioengineering. A strong growth of this developing field over the past decade has been stimulated by worldwide advocations on brain and related research. Its further growth in the short term is likely to be bottlenecked by the scientific and technological hurdles to expand the corresponding market with mature consumer products though a range of fascinating products such as the brain–computer interfaces (BCIs) are being envisioned or under diligent development, which continues catalyzing public interests in this field and promoting societal factors for the growth of this field.

Because electrical neural electrodes are the most developed approach to interface the nervous systems, and they represent the most reliable approach to build therapeutic or other application-oriented systems, learning interests in their fundamental science and engineering from the newer generation of engineers, scientists, clinicians, and entrepreneurs are ever high. Although there are a number of related books existing on the market, most of them are edited volumes unsuitable to be used as a textbook for classroom teaching. For the rest few textbooks on this relevant topic, they either do not cover this topic with sufficient depth and breadth or do not include the most significant and exciting developments over the past decade. This book is to meet such a growing demand and fill the urgent gap for a fundamental textbook. It has stemmed from and been motivated by my biannual teaching of an introductory neuroengineering course for upper-level undergraduate and graduate students from multiple departments across the Colleges of Engineering, Arts and Sciences, and Medicine at The Ohio State University.

Specifically, this textbook devotes to the focused topic of electrical cellular neural interfacing with an emphasis on quantitative modeling and analysis, while covering both conventional and emerging approaches. It is intended to be used as a textbook in a one-semester/term advanced course for upper-level undergraduate and graduate students in the department of electrical engineering, bioengineering, or biomedical engineering. It can also be used by other researchers and professionals with basic knowledge preparations on electrical circuits and frequency analyses as a self-learning material.

This book covers a breadth of neuroelectrophysiological approaches, including conventional and emerging ones. Equivalent electrical circuit modeling and analysis are used to elaborate the mechanisms of signal generation and transduction, principles of operation, analysis of signal transmission, and interpretation of resulting signals. This systematic body of knowledge on electrical neural interfacing with its featured analytical approaches is introduced stepwise with three key parts. In Part I, the properties and models of neurons and electrodes are introduced to lay the foundation for subsequent higher level modeling and analyses. In Part II, the principles of electrical neural recording are introduced, covering a wide range of important approaches in a stepwise manner. And in Part III, the principles of electrical neural stimulation are introduced, covering both the fundamental and specialized topics. Meanwhile, an introduction and an application chapter are provided to focus the topic of discussion, introduce the problems under consideration, and connect and extend the knowledge to real-world applications.

Exercise problems are provided at the end of each chapter. They are carefully designed to foster practicing and comprehending the key knowledge covered in the chapter. Instructors can use these problems for in-class exercises or homework assignments. They can also be adapted for quizzes and exams. Self-learning individuals can use these problems to evaluate and foster their learnings, as well.

Learning and understanding this systematic body of focused knowledge and skills are imperative for the students and researchers to properly design, analyze, develop, and use the associated neuroelectrophysiological approaches in neuroscience research, neuroengineering tool, and product developments, while they passionately contribute to the further growth of this exciting field of neuroengineering.

Columbus, USA   Liang Guo
March 15, 2021

Acknowledgments

This book has been inspired and motivated by my biannual teaching of an introductory neuroengineering course at The Ohio State University. My students' enthusiasm and feedback on the course materials and related neurotechnologies over the past seven years have been a major drive for me to prepare and complete a focused textbook in this exciting area. The motivation to teach some topics more clearly and thoroughly has driven me to dig into the topics further and formulate a more comprehensive and systemic understanding. This completed textbook is my best answer and gift for this passionate younger generation who is eager to dive into this exciting field of neuroengineering.

This book was written during the COVID-19 pandemic, when my normal research activities were suddenly interrupted and I had to stay at home for almost all of the time. Living an unusual life like this is not pleasant and easy for anyone. But this pandemic provides an unprecedented opportunity for us to pause and rethink the routines that we used to run ourselves in. I kept asking myself, what kind of research do I really want to pursue? What are the most important problems in my field of neuroengineering? What do I really want to do? Writing a good textbook in my field is one of the things that I had been dreaming of. Earlier, I had edited a Springer book, *Neural Interface Engineering: Linking the Physical World and the Nervous System*, which has been a popular one on the market. But I realized that an edited volume was not well focused and consistent; and moreover, it couldn't serve as a textbook for a neuroengineering course. Therefore, I finally sat down amid this pandemic to compose this textbook that I had been craving in my soul.

During this unusual pandemic, I have been fortunate to be surrounded by my two little daughters, *Stella* and *Gillian*, who are eager to read books, particularly including those books written by their daddy. I have also been consistently encouraged and supported by my wife, *Jenny*, to work on this textbook and to finish it early. With this writing journey, I am glad that my pandemic time has not been wasted in panic and wandering but has been filled with loves, supports, and some unusual accomplishments.

Integrating research and teaching is one of the educational goals of my funded National Science Foundation Faculty Early Career Development Program (CAREER) project (Grant # 1749701). Under this generous support, not only have I derived a series of high-quality research papers from my classroom teaching, but I have also taught these research results back in my classes, particularly in my introductory neuroengineering class. And now, I am taking a further step to disseminate these research and teaching results through this textbook for a broader impact to the neuroengineering curriculum worldwide.

Last, I want to say, *Life is Good*. Although we couldn't choose not to get immersed in this pandemic, we can choose how to live through it and can still make unusual life achievements in such an unusual time. After all, our lives are in our own hands.

Contents

Part IV Applications

About the Author

Dr. Liang Guo is a scholar, writer, and entrepreneur on neurotechnologies. He received the B.E. degree in biomedical engineering from Tsinghua University, Beijing in 2004 and the Ph.D. degree in bioengineering from Georgia Institute of Technology, Atlanta, GA in 2011. Subsequently, he worked as a postdoctoral associate in the Langer/Anderson Lab at Massachusetts Institute of Technology, Cambridge, MA. He started as an assistant professor of Electrical and Computer Engineering and Neuroscience in September 2013 at The Ohio State University. He has dived into the neuroengineering field since his sophomore year in college and has been passionate to adhere to this field through his Ph.D. and postdoc trainings. His primary research interests are in neural interface engineering and biocircuit engineering as applied to neuroscience and neural prosthetics. He has published over 30 journal papers and obtained five patents. Earlier, he had edited a Springer book *Neural Interface Engineering: Linking the Physical World and the Nervous System*. Among other honors, he was awarded the Defense Advanced Research Projects Agency (DARPA) Young Faculty Award in 2017, the National Science Foundation (NSF) Faculty Early Career Development (CAREER) Award in 2018, and the OSU College of Engineering Lumley Research Award in 2018.

List of Abbreviations

AC	Alternating current
aCSC	Anodic charge storage capacity
AP	Action potential
BCI	Brain–computer interface
cCIC	Cathodic charge injection capacity
cCSC	Cathodic charge storage capacity
CIC	Charge injection capacity
CMOS	Complementary metal oxide semiconductor
CSC	Charge storage capacity
CSD	Current source density
CSR	Current source resolution
CV	Cyclic voltammogram
DBS	Deep brain stimulation
DC	Direct current
eAP	Extracellular action potential
ECoG	Electrocorticography
ECM	Extracellular matrix
EDL	Electrical double layer
EEG	Electroencephalogram
eFP	extracellular field potential
EIS	Electrochemical impedance spectroscopy
epsAP	Extracellular perisomatic action potential
EPSP	Excitatory postsynaptic potential
FDA	Food and Drug Administration
FES	Functional electrical stimulation
FET	Field-effect transistor
GND	Ground
iAP	Intracellular action potential
IHP	Inner Helmholtz plane
IPSP	Inhibitory postsynaptic potential
ITO	Indium tin oxide

KHFAC	Kilohertz frequency alternating current
LFP	Local field potential
LTI	Linear time-invariant
MEA	Microelectrode array
MOSFET	Metal oxide semiconductor field-effect transistor
NITC	Neural information throughput capacity
OHP	Outer Helmholtz plane
PBS	Phosphate-buffered saline
PEDOT	Poly(3,4-ethylenedioxythiophene)
PPy	Polypyrrole
PPy/pTS	Polypyrrole/p-toluenesulfonate
RF	Radio frequency
RMS	Root-mean-square
SNR	Signal-to-noise ratio
SUA	Spatial unit of analysis
TCPS	Tissue-culture polystyrene
UEA	Utah electrode array

Chapter 1
Introduction

This book is about the quantitative analyses of neuroelectrophysiological approaches using a variety of neural electrodes (Fig. 1.1), including both conventional and emerging ones. This class of fine tools has been a major driving force to develop our understanding on the dynamic functions of our nervous systems, for the phenomenon that neural communication and computation manifest characteristic electrical signals. Implantable neural electrodes in the form of microelectrode arrays (MEAs, Fig. 1.1a, b) have also been the most advanced neural interfaces in developing state-of-the-art brain–computer interfaces (BCIs) for the direct control of machines or robots by mind. Thus, neural interfaces, and specifically neural electrodes, are the key to building such biotic-abiotic symbiotic systems.

1.1 Neural Electrodes

Based on functionality, neural electrodes are classified as recording or stimulating electrodes. Both recording and stimulating neural electrodes need to be placed in the vicinity of the neural target. Neural recording electrodes are essentially a type of electric potential sensor made of solid-state materials to capture the neural potential intracellularly, extracellularly or even in or on a neural tissue. Specifically, in this book, as we only consider neural interfacing at the cellular level, these neural electrodes are in the micro or even nano scale. Figure 1.1 depicts some examples of neural recording microelectrodes. Neural recording microelectrodes only work in a solution environment where the neuronal sources reside. Almost all neural microelectrodes work as a passive capacitor to capture a passing-by ionic current in the form of ion redistribution on the solution side of the electrode (for the detailed mechanism, please refer to Chap. 3). In the meantime, on the solid side of the electrode, electrons redistribute complementarily to transmit the electrical signal into the amplifier circuit.

© The Author(s), under exclusive license to Springer Nature Switzerland AG 2022
L. Guo, *Principles of Electrical Neural Interfacing*,
https://doi.org/10.1007/978-3-030-77677-0_1

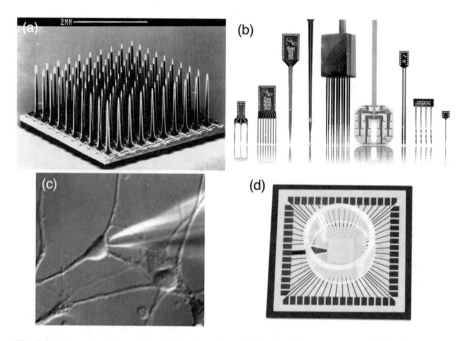

Fig. 1.1 Examples of neural recording electrodes. (**a**) The Utah electrode array (UEA), first in class cleared by the U.S. Food and Drug Administration (FDA) for human implantation. Reproduced with permission from (Kim et al. 2006). Copyright © 2006 Biomedical Engineering Society. (**b**) The NeuroNexus probes, versatile tools for neuroscience research. Image courtesy of NeuroNexus. (**c**) Patch clamp electrode in whole-cell recording configuration, most accurate tool for intracellular neural recording. Reproduced with permission from (Sigworth and Klemic 2005). Copyright © 2005 IEEE. (**d**) Culture-dish MEA, most popular tool for studying neural networks *in vitro*. Image courtesy of Multi Channel Systems MCS GmbH

Neural stimulating electrodes are transducers frequently made of noble metals (e.g., platinum (Pt)), metal oxides (e.g., iridium oxides) or conducting polymers (e.g., poly(3,4-ethylenedioxythiophene) (PEDOT)) to inject an electric current into the neural tissue, extracellular or intracellular space. They convert an electron-based current in the electronics and electrode into an ion-based current in the solution primarily through interfacial faradaic reactions (for the detailed mechanism, please refer to Chap. 4). This ionic current charges or discharges the capacitive neuronal membrane, causing corresponding transmembrane voltage changes, and may lead to the onset of an action potential (AP) if the transmembrane voltage is depolarized to its threshold. Neural stimulating electrodes may also be in the micro scale, but frequently are made larger for the purpose of injecting a larger current. Figure 1.2 depicts some examples of neural stimulating electrodes.

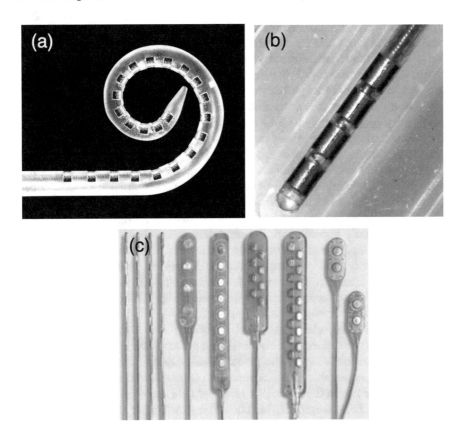

Fig. 1.2 Examples of neural stimulating electrodes. (**a**) Cochlear implant electrodes. Reproduced with permission from (Roland Jr. 2005). Copyright © 2005 The American Laryngological, Rhinological and Otological Society, Inc. (**b**) Deep brain stimulation (DBS) electrodes. Reproduced with permission from (Husch et al. 2018). Copyright © 2017 The Authors. Published by Elsevier Inc. (**c**) Spinal cord stimulation electrodes. Reproduced with permission from (Epstein and Palmieri 2012). Copyright © 2012 Mount Sinai School of Medicine

1.2   Advantages and Limitations of Electrical Neural Interfacing

Using electrodes to record or stimulate neurons is not the only approach for neural interfacing; and there are a variety of other physical means to transmit signals between the nervous system and an engineered system (Guo 2020a), for example, optical neural recording using voltage-sensitive dyes and optical neural stimulation using optogenetics. However, electrical neural interfacing does have a number of prominent advantages over other approaches, making it the obvious choice in many applications, particularly in implanted ones. First, simplicity and reliability. Being simple in structure or configuration infers operational reliability. Electrical neural interfacing requires the placements of a neural electrode in the vicinity of the neural target no matter it is for recording or stimulating and a reference or counter electrode some distance away. Once the closed circuit is configured properly, the signal can be read or written. Nowadays, electrical neural interfacing devices and systems become cheaper and easier-to-use tools in the field. Once implanted, in the short term without being bothered by the foreign body reactions, neural electrodes can operate faithfully. Second, in many cases, the recorded signals can be directly traced to certain neural activities, e.g., APs, with little ambiguity. Third, high temporal resolution. An electrical neural recording system can easily exceed the neural recording requirements (Guo 2020b), which other approaches, such as optical recording, struggle to meet. For stimulation, short current or voltage pulses in the microsecond scale can be easily produced. Fourth, parallelism. Hundreds or even thousands of independent recording channels can be implemented using state-of-the-art electronics, providing rich information on the underlying neural population. Fifth, small form factor suitable for chronic implantation. An electronic recording or stimulating system can be implemented as a standalone fully implantable device which can be battery powered internally and wireless controlled externally using radio-frequency (RF) signals. The *in vivo* operation of such implanted devices can range from months to years, depending on the specific application and design. This capability can hardly be approached by other physical interfacing modalities.

On the other hand, electrical neural interfacing also faces its own intrinsic limitations. First and foremost, contact interfacing. Due to the short range of electric fields, neural electrodes have to be placed in the vicinity of the neural target and ideally as close as possible. This requirement causes a number of problems: (1) a higher spatial resolution (i.e., a smaller receptive field during recording or a smaller activating field during stimulating) is correlated to a smaller electrode size, which can pose a technical constraint due to fabrication or accessibility challenges. (2) Invasiveness of the approach and spatial accommodation problem of the implant and accessories in *in vivo* applications. (3) Poor chronic performances due to foreign body reactions occurring surrounding the implant. Second, as almost all in vivo applications involve extracellular interfacing, due to the high packing density of neurons and the diffusiveness of electric fields, it becomes challenging to resolve the neuronal sources from recordings and to selectively activate certain neurons during stimulation. Third, as electrical neural stimulation is in an unnatural form, neuronal and tissue damages can result from improper or long-term stimulation.

Certainly, there is still much room for further developments in this area. Nonetheless, intensive research is being conducted to push the frontiers of this area and achieve a proper balance between the pros and cons for a specific application.

1.3 Problems of Focus in This Book

For this broad class of neuroelectrophysiological approaches, the problems under consideration in this book include: (1) the universal operating mechanisms of almost all neuroelectrophysiological approaches, (2) proper configuration of each approach, and (3) proper interpretation of the resulting signals. Efforts are made to both extract the universal principles underlying this common class of approaches and discern the unique properties of individual approach. The knowledge developed will help to promote understanding on the operating mechanisms of this class of tools, inform their proper usage and development, and improve the proper interpretation of the associated signals and phenomena.

1.4 Featured Approach of Analysis

To address the above problems, in this book, I use equivalent electrical circuit modeling and signal analysis to unravel the functioning mechanisms and principles and provide sound interpretations to the associated signals and phenomena. Both the neurons and electrode systems can be modeled as equivalent electrical circuits to coalesce into a whole circuit for analyzing the dynamic input–output signal relationships, as well as the intermediate signal transmissions at certain physical stages of the circuit. This book aims to derive analytical solutions to these equivalent circuits, which can offer clear and complete mechanistic insights to the underlying biophysics.

Exercises

1.1 Please answer:

    (a) What are neural electrodes?

    (b) Why do we need neural electrodes?

    (c) What are the advantages and disadvantages of interfacing the nervous systems using electrodes?

1.2 Search on the internet and find some neural electrodes. What applications are they used for? Are they recording or stimulating electrodes?

1.3 Search on the internet and find some recorded neural signals. Where were these signals obtained? What kind of neural electrodes were used to acquire these signals?

1.4 Please name some neuroelectrophysiological techniques and explain what they are used for.

1.5 Search on the internet and find out what other means are used to interface the nervous systems. Please consider both recording (reading) and stimulating (writing) technologies.

References

Epstein LJ, Palmieri M (2012) Managing chronic pain with spinal cord stimulation. Mt Sinai J Med 79 (1):123-132. doi:https://doi.org/10.1002/msj.21289

Guo L (2020a) Neural Interface Engineering: Linking the Physical World and the Nervous System. Springer Nature,

Guo L (2020b) Principles of functional neural mapping using an intracortical ultra-density micro-electrode array (ultra-density MEA). J Neural Eng 17. doi:https://doi.org/10.1088/1741-2552/ab8fc5

Husch A, M VP, Gemmar P, Goncalves J, Hertel F (2018) PaCER—a fully automated method for electrode trajectory and contact reconstruction in deep brain stimulation. Neuroimage Clin 17:80-89. doi:https://doi.org/10.1016/j.nicl.2017.10.004

Kim SJ, Manyam SC, Warren DJ, Normann RA (2006) Electrophysiological mapping of cat primary auditory cortex with multielectrode arrays. Annals of biomedical engineering 34 (2):300-309. doi:https://doi.org/10.1007/s10439-005-9037-9

Roland JT, Jr. (2005) A model for cochlear implant electrode insertion and force evaluation: results with a new electrode design and insertion technique. Laryngoscope 115 (8):1325-1339. doi: https://doi.org/10.1097/01.mlg.0000167993.05007.35

Sigworth FJ, Klemic KG (2005) Microchip technology in ion-channel research. IEEE transactions on nanobioscience 4 (1):121-127

Part I
Properties and Models of Neurons and Electrodes

Chapter 2
Equivalent Circuit Models of Neurons

In order to analyze the electrical signal transmission from a neuron to electronics or vice versa, we need to derive an overall electrical circuit to include both the electronics and the neuron. Thus, biological neurons to be interfaced by electronics need to be modeled using equivalent electrical circuit parameters. In this chapter, we only consider the neuronal soma and idealize it as a sphere to drive the following models with lumped electrical parameters. Modeling of the fine axons and dendrites is not our focus in this book unless where needed, because almost all of the neuroelectrophysiological techniques to be discussed in this book are not being used to primarily record or stimulate these fine processes.

2.1 The Classic Parallel-Conductance Model

Figure 2.1 depicts the classic parallel-conductance model (Johnston and Wu 1995) of the neuronal plasma membrane, which accounts for the more universal situation. First, the lipid bilayer plasma membrane is modeled as a capacitor C_m, with a specific membrane capacitance of 0.01 pF/μm^2 for a typical cell (Johnston and Wu 1995). Second, three types of ion channels are added in parallel to account for the different transmembrane ionic currents with corresponding lumped ion-channel conductances g_{Na}, g_K, and g_{Cl} and equilibrium potentials E_{Na}, E_K, and E_{Cl}. Opening probabilities of the Na$^+$ and K$^+$ channels depend on the transmembrane voltage $v_m(t)$, which is defined as the electric potential deference between the inside and outside (i.e.,

© The Author(s), under exclusive license to Springer Nature Switzerland AG 2022
L. Guo, *Principles of Electrical Neural Interfacing*,
https://doi.org/10.1007/978-3-030-77677-0_2

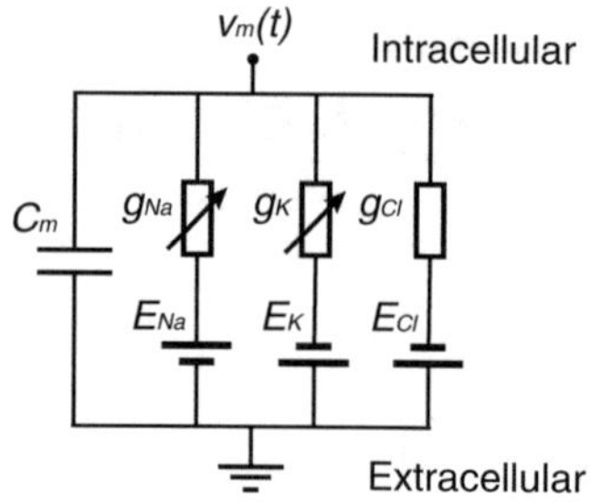

Fig. 2.1 The classic parallel-conductance model of the neuronal plasma membrane

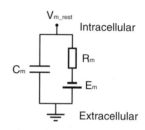

Fig. 2.2 DC model of the neuronal membrane

$v_m(t) = v_{inside}(t) - v_{outside}(t))$, and thus the channel conductances g_{Na} and g_K are functions of the $v_m(t)$. The leaking conductance g_{Cl} is constant, independent to $v_m(t)$.

In this book's analyses, we do not use this parallel-conductance model directly, but it is worth knowing that the other neuronal models described in this chapter are customized variants to this classic model for specific analysis purposes. During rest or equilibrium, g_{Na} and g_K are so small that these two parallel paths can be treated as an open circuit, and the model reduces to the DC model in Sect. 2.2. During subthreshold membrane voltage changes, the g_{Na} and g_K paths are still open circuit; E_{Cl} is shorted to the AC $v_m(t)$; and the model reduces to the AC model in Sect. 2.3. During AP, E_{Na}, E_K and E_{Cl} are all shorted to the AC $v_m(t)$, and the three conductances are superimposed to form a single transmembrane current source with an equivalent internal resistance, as shown in the AC model in Sect. 2.4.

2.2   Neuronal Model for DC Analysis

DC analysis is used during intracellular recording of the resting membrane potential V_{m_rest} (see Chap. 5). Figure 2.2 depicts the DC model of the neuronal membrane. The neuronal membrane is simply modeled as a capacitor C_m in parallel with a resistor R_m. The capacitor C_m accounts for the capacitance of the lipid bilayer plasma membrane, with a specific membrane capacitance of 0.01 pF/μm^2 for a typical cell (Johnston and Wu 1995). The resistor R_m accounts for the membrane resistance through the leaking Cl$^-$ (and K$^+$) ion channels. At rest, the Nernst potential $V_{m\_rest} = -E_m$, so that the net transmembrane current $I_m = 0$. For example, typical R_m and C_m values for HEK 293 (human embryonic kidney 293) cells with an average diameter of 13μm are between 150 and 600 MΩ and 10 and 30 pF, respectively (Robinson et al. 2012). These two parameters stay relatively constant and do not vary with the transmembrane voltage changes.

2.3   Neuronal Models for AC Analysis

AC analyses are of a higher interest, as most time we are recording the APs of neurons extracellularly and we need to understand what we are recording. The AC neuronal models provide us with the proper tools to derive the relationship between the transmembrane voltage $v_m(t)$, including both subthreshold changes and APs, and our recording (see Chap. 6).

2.3.1   Neuronal Model for Analyzing Subthreshold Transmembrane Voltage Changes

During subthreshold membrane depolarization (e.g., excitatory postsynaptic potential (EPSP)) or hyperpolarization (e.g., inhibitory postsynaptic potential (IPSP)), the neuronal plasma membrane still behaves passively and is represented by the parallel capacitor and resistor model as depicted in Fig. 2.3. Although the values of these two lumped parameters still stay the same, the membrane capacitor C_m plays a major role in these AC analyses, as opposed to merely holding the intracellular voltage in the DC model. The frequency-domain representation of the transmembrane voltage $V_{msub}(s)$ signifies the difference from the DC model in Fig. 2.2.

Fig. 2.3 Neuronal model
for analyzing subthreshold
transmembrane voltage
changes

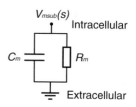

Fig. 2.4 Neuronal model
for analyzing suprathreshold
transmembrane voltage
changes

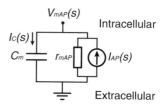

2.3.2   Neuronal Model for analyzing Suprathreshold Transmembrane Voltage Changes (APs)

During AP, the neuronal plasma membrane transforms to a self-perpetuated electromotive force, i.e., a current source, which is instantiated by the opened voltage-sensitive Na^+ and K^+ channels as well as the leaking Cl^- channels in the membrane. A lumped current $I_{AP}(s) = I_{Na}(s) - I_K(s) - I_{Cl}(s)$ is used to represent this current source with a defined direction pointing from the outside to the inside (Fig. 2.4). The parallel ion channel resistances are also lumped and treated as the internal resistance r_{mAP} of the current source. The membrane capacitor C_m stays the same, whereas the r_{mAP} is different from the R_m in Fig. 2.3.

2.3.3   Virtual Capacitive Current $I_C(s)$

There is an interesting phenomenon in the above AC model during AP, i.e., the current $I_C(s)$ flowing from the inside to the outside across the membrane capacitor is a *virtual capacitive current*. It should be noted that this concept of virtual capacitive current is only a property of our oversimplified monopole source model and does not have a real physical representation. Please see Sect. 2.4 below for an explanation on the monopole source model.

In the above AC model during AP, the following virtual capacitive current concept holds (Guo 2020). For an imaginary neuron suspended in an electrolyte and referring to Fig. 2.4, during the AP, there is an outward membrane capacitive current $I_C(s) = -I_{AP}(s)$ to close the circuit. As the neuron has uniform current densities (current per unit membrane area) across its entire membrane surface, the capacitive current balances the inward Na^+ current during the depolarization phase and the outward K^+ current during repolarization. A close scrutinization on this capacitive current from the biophysical aspect of membrane depolarization and repolarization makes us aware that it does not cross the membrane and does not flow into the extracellular space. Its existence is merely a passive consequence of the discharging or recharging of the transmembrane voltage $V_{mAP}(s)$ by the $I_{Na}(s)$ or $I_K(s)$ according to $I_C(s) = C_m \cdot sV_{mAP}(s)$. Take the discharging phase by $I_{Na}(s)$ as an example. At rest, the membrane is negatively charged inside with anions accumulated on the inner membrane surface and cations on the outer membrane surface. When the subthreshold depolarization reaches the AP threshold, noticeable Na^+ ions start to flow through opened Na^+ ion channels across the membrane from the outside. The transportation of one Na^+ ion from the outside to the inside, where it "cancels out" an anion, depolarizes (reduces) the $V_{mAP}(s)$, which requires removal of one charge from both sides (a positive charge from the outside and a negative charge from the inside) of the membrane capacitor C_m according to $Q = C_m V_{mAP}(s)$. Interestingly, this process automatically meets this requirement of charge pair removal without an actual current flowing to either the extracellular or the intracellular space. This conclusion can be similarly extended to the recharging phase where the K^+ current takes effect. Thus, this type of transmembrane capacitive current $I_C(s)$ is termed as a "virtual" current. Therefore, the consequence in our AC model during AP in Fig. 2.4 is that a net current $I_{AP}(s)$ flows into the cell and the cations stay there; meanwhile, the same current flowing into the extracellular space is denoted as $-I_{AP}(s)$, which is the key to derive the relationships between the intracellular AP (iAP) and extracellular AP (eAP) in Chaps. 6 and 9.

2.4 Monopole Current Source

In our above four neuronal models, we idealize the neuronal soma as a sphere with uniform membrane properties, which can be modeled as parallel lumped electrical parameters. Such a concept automatically assumes the neuron as a monopole current source, and the virtual capacitive current in Sect. 2.3.3 above is a direct consequence of this assumption. However, a real neuron, e.g., a pyramidal neuron in the cortical Layer V, can be more accurately represented as a dipole or even multipole current source (Delgado Ruz and Schultz 2014). Taking the dipole current source model as an example, when, in Fig. 2.4, the AP current $I_{AP}(s)$ enters the neuronal soma, rather than creating the virtual capacitive current with the cations staying on the inner side of the membrane, the ionic $I_{AP}(s)$ spreads laterally along the inner membrane and, in a pyramidal neuron, flows to and exits capacitively from the apical dendrite membranes into the extracellular space. Meanwhile, in the extracellular space along the neuronal membrane, a lateral ionic current flows in the opposite direction from the apical dendrite to the soma to supply the $I_{AP}(s)$ that enters the soma. Thus, a dipole current source model captures such a real physical phenomenon faithfully.

However, because the current densities in the membranes of apical dendrite are much smaller than that in the soma membrane, typical extracellular neural micro-electrodes cannot detect the currents from the apical dendrite above the electrical recording noise floor, and what is showing in the recording is only the effect of the perisomatic current, which approximates to the result of a monopole current source. Additionally, when the observation point, i.e., the recording site, is relatively far away from the soma, the monopole current source model can also faithfully capture the extracellular electric field generated by a neuron (Delgado Ruz and Schultz 2014). Thus, we adopt the much simpler monopole current source models above to derive our intuitive analytical solutions in subsequent chapters.

Exercises

2.1 Consider the intracellular action potential,

   (a) Sketch the Parallel-Conductance Model of the neuronal membrane and label it properly.
   (b) Which ionic currents are involved during the AP process, including the subthreshold depolarization phase?
   (c) How do these currents produce the profile of the AP?
   (d) Derive the transmembrane voltage $V_m(s)$ from these currents.

2.2 Consider the virtual capacitive current,

   (a) Explain the mechanism;
   (b) Why is it a direct result of the monopole current source model?

References

Delgado Ruz I, Schultz SR (2014) Localising and classifying neurons from high density MEA recordings. J Neurosci Methods 233:115-128. Doi: https://doi.org/10.1016/j.jneumeth.2014.05.037

Guo L (2020) Perspectives on electrical neural recording: a revisit to the fundamental concepts. J Neural Eng 17 (1):013001. Doi: https://doi.org/10.1088/1741-2552/ab702f

Johnston D, Wu SM-S (1995) Foundations of cellular neurophysiology. MIT Press, Cambridge, Mass.

Robinson JT, Jorgolli M, Shalek AK, Yoon MH, Gertner RS, Park H (2012) Vertical nanowire electrode arrays as a scalable platform for intracellular interfacing to neuronal circuits. Nature nanotechnology 7 (3):180-184

Chapter 3
Recording Electrodes

To detect the electric potentials produced by neurons in a solution environment, we use a solid-state sensor called electrode made of highly selected materials. A solid-state neural recording electrode functions as an electrical transducer in an electrolytic solution to convert a small portion of a passing-by ionic current in the solution into an electronic current flowing into the electronic amplifier. Understanding how a neural recording electrode captures the neuronal potential in an electrolytic solution is critical to proper interpretation of the recorded signal, as well as rational design of electrodes and proper selection of the matching signal amplifier. This topic starts with the physical phase boundary at the electrode–electrolyte interface in an electrochemical cell, before the entire recording circuit can be established for analysis.

3.1 Electrode–Electrolyte Interface

3.1.1 The Universal Electrode–Electrolyte Phase Boundary

Referring to Fig. 3.1, a phase transition boundary is developed when a solid (conductor, semiconductor, or insulator) is brought in contact with a salt (electrolytic) solution. During the initial transition process, atoms in the solid surface may lose electrons to the solution and become dissolved, or ions in the electrolyte may gain electrons from the solid and become deposited/adsorbed on its surface. Over a short period of time when this solid–electrolyte interface achieves a thermodynamic equilibrium, on the one hand, a thin charge (electron or hole) layer forms immediately underneath the solid surface; and on the other hand, complementary ionic charges spatially coalesce in the electrolytic solution to yield a concentration gradient and a charge gradient (together called an electrochemical gradient) conventionally described by the so-called *electrical double layer* (EDL). As a result, an electrostatic field is formed starting from the immediate solid surface and extending into the electrolytic solution, and a characteristic electrode potential (*i.e., the surface*

© The Author(s), under exclusive license to Springer Nature Switzerland AG 2022
L. Guo, *Principles of Electrical Neural Interfacing*,
https://doi.org/10.1007/978-3-030-77677-0_3

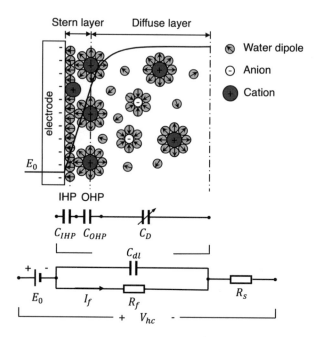

Fig. 3.1 Electrode–electrolyte interface at equilibrium and its equivalent electrical circuit model. The EDL comprises the Stern layer (including the IHP and OHP) and the diffuse layer. In the Stern layer, the magnitude of electric potential decreases linearly with distance, while in the diffuse layer, it decreases exponentially. A series of three capacitors models the IHP, OHP, and diffuse layer, respectively, and is usually lumped together as C_{dl} in the overall equivalent circuit model. E_0: surface potential of the electrode, R_f: Faradaic resistance through redox reactions, I_f: Faradaic current, R_s: spreading (or series) resistance of the bulk electrolyte, and V_{hc}: half-cell potential. Reproduced with permission from (Guo 2020a). Copyright © 2020 IOP Publishing Ltd

potential E_0) is established, which is associated with the solid's material type. The Bockris/Devanathan/Müller model describes this EDL as a cascade of three regions (Massobrio et al. 2016; Horch and Dhillon 2004; Huang et al. 2015; Valov 2017; Kampouris et al. 2015; Pilon et al. 2015): (1) an *inner Helmholtz plane* (IHP) formed by a layer of adsorbed water molecules with highly oriented dipole moments on the solid's surface, (2) an intermediate *outer Helmholtz plane* (OHP) formed by solvated (hydrated, in aqueous solutions) ions, and (3) a *Gouy–Chapman diffuse layer* beyond the OHP with an ionic charge density decaying outward due to increasing thermal motions of ionic species in the solution. The IHP and OHP together are summarized as the *Stern layer*. Additionally, some specifically adsorbed, and partially solvated ions can also appear in the IHP, as proposed by D. C. Grahame.

The electrostatic field in the EDL has two regions: In the Stern layer, the magnitude of potential decreases linearly with distance from the solid's surface, and the boundary potential at the outer edge of the OHP is referred as the *Stern potential*. In the diffuse layer, the magnitude of potential decreases exponentially. The thickness of this diffuse layer is called the *Debye length* (Horch and Dhillon

2004). In aqueous solutions, it is typically on the scale of a few nanometers and decreases with increasing ion concentration.

The electrical properties of this EDL are conventionally modeled as a series of three capacitors, representing the spatial ionic charge distributions in the IHP, OHP, and diffuse layer, respectively (Massobrio et al. 2016). The full equivalent electrical circuit model also includes a parallel Faradaic resistance R_f, the surface potential E_0 of the electrode, and the series resistance R_s of the bulk electrolyte, as depicted in Fig. 3.1. The R_f further consists of two components (not depicted): the crossover electron transfer that is proportional to the exchange current density and the mass transport phenomena that are modeled with the so-called *Warburg impedance* (also called the *constant phase element*) (Horch and Dhillon 2004). In the physiological frequency range, the bulk electrolytic solution is considered as a pure resistor R_s (Horch and Dhillon 2004; Chaparro et al. 2016). For example, phosphate buffered saline (PBS) has a resistivity of ~0.09 $\Omega \cdot$m at room temperature (Chaparro et al. 2016), which give R_s a value of around 1 kΩ for a microelectrode immersed in the center of a 10 cm diameter Petri dish.

3.1.2 Neural Recording Electrodes are Capacitive

During neural recording, electrons do not cross over the electrode–electrolyte interface because the electric potentials generated by neurons are too low to initiate any Faradaic reactions in the nearby EDL of the electrode. Thus, the R_f path in Fig. 3.1 can be eliminated, and the electrode–electrolyte interface can simply be modeled as a pure capacitor C_{dl}. Both noble-metal electrodes such as those made of platinum or gold, and conducting polymer electrodes, such as those made of polypyrrole (PPy), function in this way.

However, the electrochemical impedance spectra of electrodes made of noble metals and conducting polymers appear distinctly in the neural signal spectrum (see Fig. 3.4). Special care needs to be taken to properly interpret these impedance spectra, because during electrode impedance measurement (see Sect. 3.4) a millivolt cosinusoidal signal is injected from the electrode to the electrolyte, which makes the electrode functions differently from the neural recording scenario where the extra-cellular APs (eAPs) only have amplitudes on the scale of hundreds of microvolts or lower, given that almost all of these electrodes are used for extracellular recording. Under such a microvolt scale, the EDL C_{dl} of PPy electrodes is sufficient to transfer the current and it is unlikely for the redox reaction to take place in the polymer's backbone.

The electrochemical impedance spectra of noble-metal electrodes show a dominating capacitive property at low frequencies (see Fig. 3.4 for an example), whereas those of conducting polymer electrodes show a strong resistive property across the spectrum. During electrochemical impedance spectroscopy (EIS) with a very small cosinusoidal testing voltage (e.g., a 2.5 mV amplitude), there is no Faradaic reaction occurring in the EDL of noble-metal electrodes, which thus function as a pure capacitor C_{dl}, and the electrochemical impedance spectrum represents the electrode's recording impedance faithfully. In contrast, even with the small testing current, a reversible redox reaction still happens in the conducting polymer's backbone to facilitate the current delivery ionically between the polymer and electrolyte (Guo 2016). However, no electrons are released from the backbone into the electrolyte, thus no "real" Faradaic reactions happen in the EDL, and the conducting polymer electrode functions as a pseudo-capacitor. But the R_f path in Fig. 3.1 always exists in such EIS testing, which is not the case during actual neural recording as stated above. As a result, the electrochemical impedance spectrum of a conducting polymer electrode does not represent the electrode's recording impedance.

3.2 Non-Redox Electrochemical Cell for Neural Recording

Figure 3.2 depicts the complete electrode configuration for an extracellular neural recording, which is a non-redox electrochemical cell, because no redox reactions take place in the electrochemical cell. The recording electrode only passively experiences a passing-by ionic current $-i_{AP}(t)$ ($i_{AP}(t)$ is defined as the inward current into the neuron during AP, see Sect. 2.3.2), and capacitively absorbs a very small portion of it to produce a voltage reading across the input terminals of the differential amplifier. The reference electrode is placed at a quiescent site where no ionic current flows by and thus is virtually shorted to the ground (*GND*) through the intermediate solution. The *GND* electrode is placed at a site far away from the recording and reference electrodes and sets the zero potential. The recording and reference electrodes are connected to the two input terminals of a different amplifier, respectively. These two electrodes should be of the same properties, including material composition and geometry. This non-redox electrochemical cell cannot be explained by the

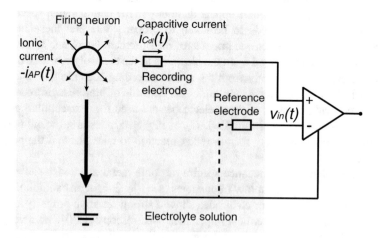

Fig. 3.2 Non-redox electrochemical cell for neural recording

conventional galvanic or electrolytic electrochemical cell, both of which inevitably involve a matching pair of redox half-reactions (Guo 2020a). Therefore, a new theory has been developed to explain the recording mechanism (Guo 2020a), as detailed in Sect. 3.3 below.

3.3 The Complete Neural Recording Circuit

To elucidate the mechanism of electrical recording in an electrolyte, we need to construct the full electrical circuit of the recording system as shown in Fig. 3.3, including equivalent subcircuits of the neuronal current source (see Chap. 2) and the electrode–electrolyte interfaces. As an example, in such a comprehensive circuit, we assume the recording is performed extracellularly, and thus the diverging ionic current $I_X(s)$ produced by the neuron at the extracellular observation Point X flows to the signal ground GND placed at the infinity only through the resistive path R_s in the absence of the recording circuit. When the recording electrode is placed at Point X and connected to the amplifier circuit, the electrode and amplifier circuit provide an additional parallel path for $I_X(s)$ to flow to the GND. Our analyses below center around this circuit architecture. This overall circuit architecture also applies to intracellular configurations with appropriate modifications (see Chap. 5).

In the electrolyte, the flow of an ionic current establishes an electric field that can exert a force on electric charges in this conductor. This force is what we measure in the form of a potential difference (Johnston and Wu 1995). In Fig. 3.3, with the signal ground GND placed far away and in the absence of the electrode, the electrical potential at Point X in the electrolyte is determined by *Ohm's law* through $V_X(s) = R_s I_X(s)$. Therefore, for a given $I_X(s)$, the field potential $V_X(s)$ at that point depends on the effective impedance of the leaking path for $I_X(s)$ to flow to the GND (Hai et al. 2009). This perspective is supported by, for example, the finding that the lumen of a microchannel can effectively boost the amplitude of extracellularly recorded APs (FitzGerald et al. 2009, 2008), because of the substantial increase of R_s in the microchannel. It is worth noting that, by applying the *Thevenin's Theorem*, the current source $I_X(s)$ in Fig. 3.3b can be converted to an equivalent voltage source $V_X(s) = R_s I_X(s)$ with the same internal source resistance R_s, as shown in Fig. 3.3c where the voltage $V_X(s)$ is assumed at the infinity but actually located at Point X during recording (Fig. 3.3b). When the electrode is present, $Z'_e = Z_e + R_s$ (Fig. 3.3c) is the *conventional electrode impedance* measured in an open-field physiologically relevant electrolyte (such as PBS) using EIS, assuming Faradaic processes are not involved during the measurement (*i.e.*, the Faradaic resistance $R_f = \infty$). Thus, R_s appears to belong to the signal source, which generates $V_X(s)$ from $I_X(s)$.

In Fig. 3.3a, an AC amplifier is used, in contrast to the DC amplifier used in conventional intracellular recording under current-clamp mode (see Chap. 5), where a finite faradic resistance exists in the electrode's EDL (Fig. 3.1) (Hai et al. 2010a, b). Considering the AC nature of neural signals, since no current flows in the reference

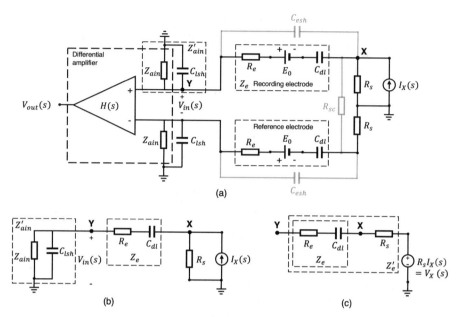

Fig. 3.3 Equivalent electrical circuit formed between the recording electrode and amplifier input. The eAP current propagating to the edge (Point X) of EDL of the recording electrode is represented as a current source $I_X(s)$. The recording and reference electrodes are assumed to be made of the same materials and geometric dimensions, thus having the same electrical properties. (**a**) The full equivalent circuit. (**b**) The reorganized voltage-divider circuit in the context of AC signal transmission. (**c**) The current source can be converted to an equivalent voltage source using *Thevenin's Theorem*, where R_s appears as the internal resistance of both signal sources. R_s: spreading resistance, R_{sc}: potential resistive coupling through the electrolytic solution between the recording and reference electrodes (if the two are placed far away so that $R_{sc} = \infty$, this resistive coupling can be ignored), C_{dl}: EDL capacitance, E_0: electrode surface potential (in an AC circuit, this electromotive force is equivalent to a short wire), R_e: resistance of the electrode material, C_{esh}: shunt capacitance across the insulation between the electrode shaft and surrounding electrolyte (if the electrode is properly designed and fabricated, it can be ignored), Z_e: the effective electrode recording impedance, C_{lsh}: cumulative shunt capacitance along the lead and connector between the electrode and amplifier input terminal, Z_{ain}: input impedance of the amplifier, Z'_{ain}: the effective input impedance of the amplifier, $V_{in}(s)$: voltage appearing across the input terminals of the differential amplifier, $V_X(s)$: the potential at Point X in the absence of the recording electrode, Z'_e: the conventional electrode impedance measured in an open-field physiologically relevant electrolyte using EIS. The GND electrode is placed far away. Adapted with permission from (Guo 2020a). Copyright © 2020 IOP Publishing Ltd

electrode path, the negative terminal of the amplifier is shorted to the *GND*, and the full circuit in Fig. 3.3a is equivalent to the one in Fig. 3.3b, which is a current splitter at Point X combined with a voltage divider at Point Y. The voltage $V_{in}(s)$ appearing across the differential amplifier's input terminals at Point Y is related to the field potential $V_X(s)$ at Point X through Eqs. (3.1)–(3.3):

$$V_{in}(s) = \frac{Z'_{ain}}{Z_e + Z'_{ain}} V'_X(s) \tag{3.1}$$

where $V'_X(s)$ is the potential at Point X in the presence of the recording electrode:

$$
\begin{aligned}
V'_X(s) &= \frac{(Z_e + Z'_{ain})R_s}{Z_e + Z'_{ain} + R_s} I_X(s) = \frac{Z_e + Z'_{ain}}{Z_e + Z'_{ain} + R_s} R_s I_X(s) \\
&= \frac{Z_e + Z'_{ain}}{Z_e + Z'_{ain} + R_s} V_X(s) = \frac{Z_e + Z'_{ain}}{Z'_e + Z'_{ain}} V_X(s)
\end{aligned} \tag{3.2}
$$

Substituting Eq. (3.2) into Eq. (3.1), we have

$$V_{in}(s) = \frac{Z'_{ain}}{Z_e + Z'_{ain} + R_s} V_X(s) \tag{3.3}$$

Therefore, for extracellular electrical recording using a free-standing point electrode that does not form a tight seal with the cell membrane, R_s (e.g., in PBS, from the outer surface of a 10μm diameter spherical neuron to GND at the infinity, $R_s = \frac{\rho}{4\pi r} = 1433\ \Omega$) is orders of magnitude smaller than the serial impedance $Z_e + Z'_{ain}$ (e.g., >10 MΩ), we thus have

$$V_{in}(s) \approx \frac{Z'_{ain}}{Z_e + Z'_{ain}} V_X(s) \tag{3.4}$$

According to Eq. (3.2), the presence of the *wired* electrode at Point X tends to slightly lower its potential, as $\left| \frac{Z_e + Z'_{ain}}{Z'_e + Z'_{ain}} \right| < 1$. If $R_s \ll \left| Z_e + Z'_{ain} \right|$, we have

$$V'_X(s) = \frac{1}{1 + \frac{R_s}{Z_e + Z'_{ain}}} V_X(s) \approx V_X(s) \tag{3.5}$$

Then, this distortion is minute and negligible. It is also noted that the presence of the microelectrode alone at Point X without providing an additional path to the GND (e.g., via the amplifier terminal, thus we can assume $Z'_{ain} = \infty$) does not perceivably affect the AC potential at Point X (please also see Exercise 3.5).

Equation (3.4) proves that the electrode recording system functions as a voltage-divider circuit (attenuating the amplitude) with a second-order dynamics (see Chap. 11, Sect. 11.1.4) (Guo 2020b) to sense the AC potential in an electrolyte where the recording electrode is placed. It is noted that the *electrode recording impedance* Z_e is lumped together with R_s to form the *conventional electrode impedance* Z'_e (Fig. 3.3c), which means conceptually the voltage source is located at the infinity where the GND used to be placed, even if the $V_X(s)$ is actually present at Point X, the outer edge of the electrode's EDL. Nonetheless, in Fig. 3.3c, the voltage drop across R_s is negligible, so that $V'_X(s) \approx V_X(s)$, as shown in Eq. (3.5). However, this sensing mechanism is only made possible by virtue of capacitive current transmission through the EDL of the electrode—thus the electrode functions as a pure capacitor (see Sect. 3.1.2). This implies that *the voltage $V_{in}(s)$ sensed by the electrode recording system is actually caused by the capacitive current that crosses the electrode's EDL and is in the form of ion redistribution on the electrolyte side.* According to this voltage-divider circuit, when an ionic current flows into the electrode's EDL (*i.e.*, cations flow into or anions flow out of the EDL), a positive voltage is recorded at Point Y; and when an ionic current flows away from the EDL (*i.e.*, cations flow out of or anions flow into the EDL), a negative voltage is recorded.

3.4 Electrode Impedance

According to Eq. (3.3), the electrode recording impedance Z_e (conventionally measured as Z'_e in an open-field physiologically relevant electrolyte) is one of the critical parameters influencing the recording signal-to-noise ratio (SNR) (Guo 2020a) and thus a must-have data for any paper that reports either the development of a new type of neural electrode or the neural recording using an electrode. However, this piece of critical information is frequently either missing or not reported properly in the literature. For example, many papers only reported the magnitude *vs.* frequency plot of the electrode impedance, whereas the phase plot was missing. As the electrode impedance is a complex function of frequency, this makes it impossible for the community to (1) develop full knowledge on characteristics of the electrode and (2) use deconvolution to recover $V_X(s)$ from $V_{in}(s)$ according to Eq. (3.3). To make things worse, many studies did not measure the electrode impedance in a proper manner, resulting in inaccurate or misleading data reported. Thus, I would like to elaborate and clarify the essential aspects on electrode impedance and its measurement.

3.4.1 Principle of Electrode Impedance Measurement

To measure its impedance, the electrode is treated as a linear time-invariant (LTI) system with a current input $I_e(j\omega)$ and a voltage output $V_e(j\omega)$. Thus, the frequency response $H_e(j\omega) = \frac{V_e(j\omega)}{I_e(j\omega)}$ of the system is the Z'_e. Note, if the input is a voltage and the output is a current, the frequency response is the admittance, the reciprocal of impedance. Either way works. For this LTI system, a complex exponential signal $e^{j\omega_1 t}$ with a unique characteristic angular frequency of ω_1 generates an output as a modulated version of the same complex exponential (Oppenheim and Willsky 1997): $e^{j\omega_1 t} \xrightarrow{H_e(j\omega)}$ $H_e(j\omega_1)e^{j\omega_1 t} = |H_e(j\omega_1)|e^{j[\omega_1 t + \angle H_e(j\omega_1)]}$, where $|H_e(j\omega_1)|$ is an even function of ω and $\angle H_e(j\omega_1)$ is an odd function (therefore, the plots only show the first quadrant). So, the strategy to measure $H_e(j\omega)$ is to construct a sequence $[\omega_1, \omega_2, \omega_3, \cdots, \omega_n]$ in the working frequency range (e.g., 0.01 Hz–100 kHz, with $\omega = 2\pi f$) of the electrode and measure the corresponding $H_e(j\omega_n)$'s one by one to reconstruct the full profile of $H_e(j\omega)$. In practice, we use a function generator to produce a current (or voltage) signal $A\cos\omega_n t$ as the testing input signal and use a frequency spectrum analyzer to analyze the corresponding output voltage (or current) signal. According to *Euler's relation*, the input current (or voltage) $A\cos\omega_n t = \frac{A}{2}e^{j\omega_n t} + \frac{A}{2}e^{-j\omega_n t}$, and thus the output voltage (or current) of the electrode system is $\frac{A}{2}H_e(j\omega_n)e^{j\omega_n t} + \frac{A}{2}H_e(-j\omega_n)e^{-j\omega_n t} = A|H_e(j\omega_n)|\cos[\omega_n t + \angle H_e(j\omega_n)]$ (Guo 2020a) (also see Exercise 3.7). Therefore, extracting the gain $|H_e(j\omega_n)|$ and phase shift $\angle H_e(j\omega_n)$ from the output will reconstruct the $H_e(j\omega_n) = |H_e(j\omega_n)|e^{j\angle H_e(j\omega_n)}$ at each ω_n. $H_e(j\omega_n)$ is conventionally presented as the magnitude and phase plots in logarithmic scales (see Fig. 3.4 for an example).

3.4.2   Method for Electrode Impedance Measurement

The measurement is usually performed using a potentiostat in a three-electrode electrochemical cell comprising a working, a counter, and a reference electrode in PBS. It can also be performed using a function generator and a spectrum analyzer in a two-electrode electrochemical cell with only a working and a counter electrode. The electrode being measured is connected to the working electrode. There are two options for the counter electrode: (A) using an electrode of exactly the same characteristics, *i.e.*, materials composition and geometry, or (B) using a very large electrode whose impedance can be ignored. The working and counter electrodes need to be placed far apart. As the circuit path being measured includes the working electrode, the PBS in between, the counter electrode, and the connecting leads (usually neglectable), the measurement $M = Z_{we} + R_s + Z_{ce} + R_{leads}$. In case (A), $Z'_e = \frac{M}{2} = Z_{we} + \frac{R_s + R_{leads}}{2} \approx Z_e + R_s$; and in case (B), $Z'_e = Z_e + R_s \approx M$. For a three-electrode electrochemical cell with an accurate setting of the reference electrode whose potential is 0 V, if the potential of the working electrode can be isolated, there is also a third case (C): the working electrode's impedance can be directly obtained with the $Z'_e = M = Z_{we} + R_s$. Selection for the amplitude A in $A\cos\omega_n t$ needs to be as small as possible, *e.g.*, with an equivalent voltage amplitude of 2.5 mV, to avoid potential initiation of Faradaic reactions in the EDL of electrode. Selection for $[\omega_1, \omega_2, \omega_3, \cdots, \omega_n]$ is usually evenly spaced in the logarithmic frequency range.

   It should be noted that the testing signal for measuring a recording electrode is different from that for a stimulating electrode (see Chap. 4). In the EDL model in Fig. 3.1 of passive recording electrodes, $R_f = \infty$ and the testing signal should not generate any Faradaic currents across the EDL. In contrast, almost all stimulating electrodes function under the Faradaic regime where R_f is finite; and thus, their measurement should reflect and emulate such a working condition.

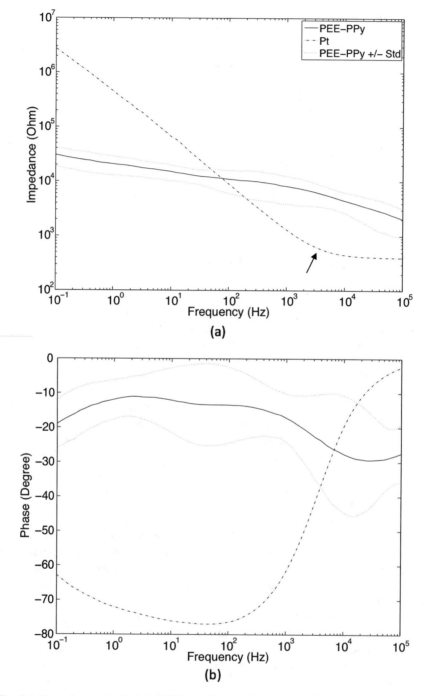

Fig. 3.4 Example magnitude (**a**) and phase (**b**) plots for the electrode impedance of a neural recording electrode array. The electrodes (1 mm diameter) and leads were directly made of the conducting polymer PPy (PEE-PPy, conductivity: 116.3 ± 7.8 S/cm). Average electrical impedance spectrum with standard deviation of four electrodes is shown. A smooth Pt disc electrode of the

3.4.3 How to Read the Impedance Plots

While the role of $|Z'_e|$ in affecting the recording SNR is generally well understood (see Eq. (3.3)), the frequent missing of the phase plot in literature reflects a lack of knowledge on the meaning of this piece of data (Oppenheim and Lim 1981). Here we use an example to help to interpret the electrode impedance plots. Referring first to Fig. 3.4a—the magnitude vs. frequency plot in log–log scales, the impedance magnitude of the Pt electrode has the well-recognizable shape: starting with a very high value at a frequency (e.g., 0.1 Hz) close to DC, the curve declines linearly (reciprocally in a linear scale) due to the decease of $\frac{1}{j\omega C_{dl}}$ with frequency until a transiting region and then becomes a flat line at high frequencies where the frequency is large enough to make $\frac{1}{j\omega C_{dl}} \approx 0$. The earlier the curve declines to the transiting point (indicated by the arrow in Fig. 3.4a), the larger the C_{dl}, as desired. The value of the flat line is actually the total resistance R_{total} in the measurement circuit, including R_s, R_e, R_{leads}, and contact resistances at joints or connections if any. For this particular Pt electrode, $R_s \approx R_{total} = 400\ \Omega$. With this information, C_{dl} can then be determined from any data point in the lower frequency range. Turning to the phase vs. frequency plot in Fig. 3.4b, at low frequencies the phase angle is close to $-90°$, as $\frac{1}{j\omega C_{dl}}$ dominates; and at high frequencies, the curve approaches to $0°$ as $\frac{1}{j\omega C_{dl}}$ zeros out and Z'_e becomes resistive.

By contrast, the impedance of the "active" PPy electrode is more complicated to interpret. The magnitude declines monotonically in the measuring frequency range without flattening at high frequencies, and the phase angle with values around $-20°$ across the frequency range shows a dominant resistive nature. This is because the PPy electrode worked primarily in reversible Faradaic processes (thus the word "active" is used) generating a pseudo capacitance to transmit charges across the frequency range even with the 5 mV test cosinusoidal signals (see Sect. 3.1.2). Thus, there was a major current path through R_f in the EDL model in Fig. 3.1, and the current path through C_{dl} was only a minor one, making the decay of the magnitude with frequency in Fig. 3.4a much slower. Furthermore, the magnitude is substantially lower at low frequencies, because the electrode surface is porous and much rougher, resulting in a much larger C_{dl}. It remains relatively high at high frequencies, because both the lead and electrode were directly made of the semiconducting PPy, thus bearing higher resistances. Additionally, PPy is known to be slow in response to current changes presumably due to Faradaic mass transport limit in the EDL (Guo et al. 2014), thus the PPy electrode's pseudo capacitance is larger at lower frequencies and declines as the frequency increases. This phenomenon is shown up in both Fig. 3.4a and b; when the pseudo capacitance is larger at lower frequencies, R_f is

---

Fig. 3.4 (continued) same geometric surface area was used as a control. The measurement configuration followed Option B (see main texts) and was the same for both electrode types. Reproduced with permission from (Guo et al. 2014). Copyright © 2013 WILEY-VCH Verlag GmbH & Co. KGaA, Weinheim

smaller, the magnitude of impedance declines slower with increasing frequencies because the resistive R_f path in the EDL dominates the current transfer with a less negative phase angle; when the pseudo capacitance becomes smaller at higher frequencies, R_f is larger (i.e., the PPy responses slower), the magnitude of impedance declines faster with increasing frequencies due to the now slightly more apparent capacitive property in the current transfer through the C_{dl} path in the EDL with a slightly more negative phase angle.

Lastly, as the characteristic frequencies of the iAP center around 1 kHz (*i.e.*, the base width of the iAP of a typical neuron is 1 ms, see Fig. 6.2), the electrode impedance is frequently measured only at 1 kHz. Although this single point of data is sufficient to calculating the C_{dl} and R_s in the EDL model of a passive neural recording electrode, the aforementioned intuitive information that can be captured from the spectrum plots is unfortunately missing. Furthermore, as will be explained in detail in Chap. 6, the characteristic frequencies of the eAP actually center around >2 kHz (see Fig. 6.2), instead, due to the first time derivative relationship to the iAP, so, for most neural microelectrodes which are used for extracellular neural recording, measuring the electrode impedance only at 1 kHz is inappropriate.

3.4.4   Methods to Reduce Electrode Impedance

Because neural recording electrodes function as a pure capacitor. To reduce the electrode impedance, we need to increase the EDL capacitance (Fig. 3.1) according to $C_{dl} = \varepsilon \frac{Area}{d}$, where ε is the absolute permittivity of dielectric in the EDL, *Area* is the equivalent electrochemical surface area of the EDL, and d is the thicknesses of the EDL. As thickness of the Stern layer and the Debye length cannot be changed in physiologically relevant electrolyte, the only option is to increase the *Area* by either roughening the smooth electrode surface (Cui and Martin 2003b) or coating it with rough and/or porous conductive materials including platinum black (Arcot Desai et al. 2010), conducting polymers (Cui and Martin 2003a; Abidian and Martin 2009) and carbon nanotubes (Keefer et al. 2008).

3.5   Summary

- The electrode–electrolyte interface of neural recording electrodes functions as a pure capacitor C_{dl}, and the extracellular field potential (eFP) that an electrode senses is at the outer edge of this C_{dl}, *i.e.*, the outer edge of its EDL.
- The voltage $V_{in}(s)$ recorded by the electrode recording system is caused by the capacitive current that crosses the electrode's EDL and is in the form of ion redistribution on the electrolyte side. When an ionic current flows into the electrode's EDL (*i.e.*, cations flow into or anions flow out of the EDL), a positive

voltage is recorded; and when an ionic current flows away from the EDL (*i.e.*, cations flow out of or anions flow into the EDL), a negative voltage is recorded.
- The presence of a wired microelectrode at a point tends to slightly lower its field potential, however, when disconnected from the amplifier's input terminal, it does not perceivably affect the AC field potential at that point.
- The electrode recording system functions as a voltage-divider circuit (attenuating the amplitude) with a second-order dynamics to sense the AC potential in an electrolyte where the recording electrode is placed.
- The electrode impedance, conventionally measured as Z'_e in an open-field physiologically relevant electrolyte, must be reported for work involving either the development of a new type of neural electrode or the neural recording using an electrode. The report should include both the magnitude and phase plots. The measurement should be performed with a proper setup (e.g., Option A, B, or C), and the testing signal should not generate any Faradaic currents across the EDL.

Note: Sections 3.1.1, 3.3, and 3.4 were adapted from (Guo 2020a) with permission.

Exercises

3.1 Consider the electrode–electrolyte interface,

(a) Describe the structure and characteristics of the EDL.
(b) What is the thickness of the EDL, and how does it vary with the concentration of the electrolyte?
(c) Why do neural recording electrodes function purely capacitively?

3.2 What is the typical configuration for extracellular neural recording? Describe the role of each components.

3.3 For neural recording electrodes,

(a) Why are low electrochemical impedances desired?
(b) How to improve the electrochemical impedance of a smooth metal electrode?
(c) Sketch an example pair of electrode impedance curves (magnitude and phase plots), and explain what information you can extract from them.

3.4 For extracellular neural recording,

(a) How is a field potential $V_X(s)$ produced in an electrolyte?
(b) Does the electrode directly sense the current or voltage? Why is the electrode conventionally regarded as a potential sensor?
(c) What does the conventional electrode impedance mean?
(d) What is the effect of the electrode recording system to the eFP? How to minimize such an effect?
(e) How is the recording $V_{in}(s)$ related to the signal $V_X(s)$? How to improve the recording quality then?
(f) Explain the physical sensing mechanism in the EDL and how $v_{in}(t)$ is generated. When is $v_{in}(t)$ positive and when is it negative?

3.5 Consider a large silicon recording probe inserted in a cortical tissue or immersed in a culture medium. Microelectrodes on the probe only occupy a small area. The probe is not electrically connected to an external recording equipment, i.e., its connecting terminals are floating in the air where the impedance is considered to be infinite. Considering the realistic situation,

  (a) Could this large probe affect the extracellular DC field potential in its surroundings in the cortical tissue or culture medium? Why and how?
  (b) Could this large probe affect the extracellular AC field potential in its surroundings in the cortical tissue or culture medium? Why and how?

3.6 For an MEA consisting of multiple smooth metal microelectrodes designed for neural recording, e.g., the NeuroNexus Probes,

  (a) Please explain why the electrochemical impedance of a recording electrode needs to be minimized.
  (b) What is the primary mechanism in neural recording?
  (c) What aspects need to consider in order to improve the performances of the MEA, and what are the methods to achieve such improvements?

3.7 Prove the principle of electrode impedance measurement, i.e., given an input current $A cos\omega_n t$, and thus the output voltage of the electrode system is $A| H_e(j\omega_n)| \cos [\omega_n t + \angle H_e(j\omega_n)]$.

3.8 For EIS,

  (a) Describe the three options for measuring the electrode impedance spectrum. If resources permit, which option produces the highest accuracy?
  (b) Why does a spectrum need to be measured, instead of only measuring at a few frequency points, e.g., at 1 kHz?
  (c) What are the special cautions when measuring the electrochemical impedance spectrum for a neural recording electrode?
  (d) How to properly represent the electrode impedance spectrum?

References

Abidian MR, Martin DC (2009) Multifunctional nanobiomaterials for neural interfaces. Advanced Functional Materials 19 (4):573-585

Arcot Desai S, Rolston JD, Guo L, Potter SM (2010) Improving impedance of implantable microwire multi-electrode arrays by ultrasonic electroplating of durable platinum black. Frontiers in neuroengineering 3:5

Chaparro CV, Herrera LV, Meléndez AM, Miranda DA (2016) Considerations on electrical impedance measurements of electrolyte solutions in a four-electrode cell. Journal of Physics (Conference Series 687):012101

Cui XY, Martin DC (2003a) Electrochemical deposition and characterization of poly (3,4-ethylenedioxythiophene) on neural microelectrode arrays. Sensor Actuat B-Chem 89 (1-2):92-102. doi: https://doi.org/10.1016/S0925-4005(02)00448-3

Cui XY, Martin DC (2003b) Fuzzy gold electrodes for lowering impedance and improving adhesion with electrodeposited conducting polymer films. Sensor Actuat a-Phys 103 (3):384-394. doi: Doi 10.1016/S0924-4247(02)00427-2

FitzGerald JJ, Lacour SP, McMahon SB, Fawcett JW (2008) Microchannels as axonal amplifiers. IEEE transactions on bio-medical engineering 55 (3):1136-1146. doi:https://doi.org/10.1109/TBME.2007.909533

FitzGerald JJ, Lacour SP, McMahon SB, Fawcett JW (2009) Microchannel electrodes for recording and stimulation: in vitro evaluation. IEEE transactions on bio-medical engineering 56 (5):1524-1534. doi:https://doi.org/10.1109/TBME.2009.2013960

Guo L (2016) Conducting Polymers as Smart Materials for Tissue Engineering. In: Smart Materials for Tissue Engineering. pp 239-268

Guo L (2020a) Perspectives on electrical neural recording: a revisit to the fundamental concepts. J Neural Eng 17 (1):013001. doi:10.1088/1741-2552/ab702f

Guo L (2020b) Principles of functional neural mapping using an intracortical ultra-density micro-electrode array (ultra-density MEA). J Neural Eng 17. doi:https://doi.org/10.1088/1741-2552/ab8fc5

Guo L, Ma M, Zhang N, Langer R, Anderson DG (2014) Stretchable polymeric multielectrode array for conformal neural interfacing. Advanced materials 26 (9):1427-1433. doi:https://doi.org/10.1002/adma.201304140

Hai A, Dormann A, Shappir J, Yitzchaik S, Bartic C, Borghs G, Langedijk JPM, Spira ME (2009) Spine-shaped gold protrusions improve the adherence and electrical coupling of neurons with the surface of micro-electronic devices. Journal of the Royal Society Interface 6 (41):1153-1165. doi:https://doi.org/10.1098/rsif.2009.0087

Hai A, Shappir J, Spira ME (2010a) In-cell recordings by extracellular microelectrodes. Nature methods 7 (3):200-U250. doi:https://doi.org/10.1038/Nmeth.1420

Hai A, Shappir J, Spira ME (2010b) Long-Term, Multisite, Parallel, In-Cell Recording and Stimulation by an Array of Extracellular Microelectrodes. Journal of neurophysiology 104 (1):559-568. doi:https://doi.org/10.1152/jn.00265.2010

Horch KW, Dhillon GS (2004) Neuroprosthetics: theory and practice. World Scientific,

Huang J, Li Z, Ge H, Zhang JB (2015) Analytical Solution to the Impedance of Electrode/Electrolyte Interface in Lithium-Ion Batteries. J Electrochem Soc 162 (13):A7037-A7048. doi:https://doi.org/10.1149/2.0081513jes

Johnston D, Wu SM-S (1995) Foundations of cellular neurophysiology. MIT Press, Cambridge, Mass.

Kampouris DK, Ji XB, Randviir EP, Banks CE (2015) A new approach for the improved interpretation of capacitance measurements for materials utilised in energy storage. Rsc Adv 5 (17):12782-12791. doi:https://doi.org/10.1039/c4ra17132b

Keefer EW, Botterman BR, Romero MI, Rossi AF, Gross GW (2008) Carbon nanotube coating improves neuronal recordings. Nature nanotechnology 3 (7):434-439. doi:https://doi.org/10.1038/nnano.2008.174

Massobrio P, Massobrio G, Martinoia S (2016) Interfacing Cultured Neurons to Microtransducers Arrays: A Review of the Neuro-Electronic Junction Models. Frontiers in Neuroscience 10. doi:https://doi.org/10.3389/fnins.2016.00282

Oppenheim AV, Lim JS (1981) The Importance of Phase in Signals. P IEEE 69 (5):529-541. doi:https://doi.org/10.1109/Proc.1981.12022

Oppenheim AV, Willsky AS (1997) Signals and Systems. 2nd Edition edn. Prentice Hall,

Pilon L, Wang HN, d'Entremont A (2015) Recent Advances in Continuum Modeling of Interfacial and Transport Phenomena in Electric Double Layer Capacitors. J Electrochem Soc 162 (5):A5158-A5178. doi:https://doi.org/10.1149/2.0211505jes

Valov I (2017) Interfacial interactions and their impact on redox-based resistive switching memories (ReRAMs). Semicond Sci Tech 32 (9). doi: https://doi.org/10.1088/1361-6641/aa78cd

Chapter 4
Stimulating Electrodes

Neural electrodes are current transducers. Recording electrodes convert an ionic current in the electrolytic solution into an electronic current in the solid-state electronics; and on the contrary, stimulating electrodes convert an electronic current in the solid-state electronics into an ionic current in the electrolytic solution. The different requirements on these two types of neural electrodes stem from the amount of current needed for conversion. In recording, a minute amount of ionic current flows into the EDL of the electrode, because (1) the neuronal current, intracellular or extracellular, is very small, and (2) the serial impedance of the electrode and amplifier input, which forms an additional parallel path, is orders of magnitude higher than the leaking impedance of the recording environment for the neuronal current to dissipate to *GND* (see Figs. 3.2 and 3.3). Thus, the physical capacitive charge transfer mechanism in the EDL suffices for such a current conversion role. In contrast, in stimulation, a large amount of electronic current is injected through the EDL of the electrode into the solution, making the physical capacitive charge transfer mechanism far incapable of meeting this demand and requiring the electro-chemical Faradaic processes to take the major role for the current conversion. Thus, while recording electrodes only need to be tailored for the physical capacitive charge transfer mechanism through C_{dl} in the EDL, stimulating electrodes need to be primarily optimized for the electrochemical Faradaic charge transfer mechanism through R_f in the EDL.

Regarding to the differences between stimulating and recording electrodes, one interesting question to ask is: can an electrode be used for both stimulation and recording? While there indeed are electrodes that can both stimulate and record, stimulating electrodes require a much higher electronic-to-ionic charge transfer capability than recording electrodes, and, generally speaking, a recording electrode cannot be used for stimulation for concerns on irreversible Faradaic reactions happening in the EDL, electrode corrosion, and biosafety. On the other hand, a stimulating electrode is generally made large in order to deliver sufficient current, making it unsuitable for recording with an adequate spatial resolution, though it can

L. Guo, *Principles of Electrical Neural Interfacing*,
https://doi.org/10.1007/978-3-030-77677-0_4

record. Other features, such as an expensive coating for a high charge injection capacity (CIC), of a stimulating electrode make it overkill for neural recording.

In this chapter, we study neural stimulating electrodes by looking into the Faradaic charge transfer processes in the EDL, analyzing the complete stimulating circuit, and characterizing the CIC.

4.1 Electrode–Electrolyte Interface

While neural recording relies on passive reception of a tiny fraction of a passing-by ionic current through the electrode's capacitive charge transfer mechanism to convert it into an electron-based current in the electronic circuit, neural stimulation intends to actively convert an electron-based current, often substantial, in the electronic circuit through the electrode into an ionic current and inject it into the electrolytic volume conductor. As shown in Fig. 3.1, there are two parallel charge transfer paths in the electrode's EDL: through the double-layer capacitor C_{dl} and through the Faradaic resistor R_f. Although the capacitive mechanism seems to be the optimal method to transfer charge into biological tissues, the charge transfer at metal stimulating electrodes mainly takes place via Faradaic processes, because the EDL's capacitive charge storage capacity (CSC), e.g., $20\mu C/cm^2$ for smooth metal electrodes, is too small to meet the required charge density. Thus, we make use of special reversible redox, i.e. Faradaic, reactions happening in the EDL coupled with a charge-balanced biphasic current stimulus waveform to achieve such a goal of charge transfer per phase; in the meanwhile, we want to avoid harmful irreversible Faradaic reactions taking place in the EDL, as well. Reversible Faradaic reactions are defined as those whose half-reaction primary products from the cathodic phase of stimulation are immobilized at the electrode's surface and can then participate in the other half-reaction during the anodic phase of restoration, or vice versa. In contrast, irreversible Faradaic reactions are those whose half-reaction products defuse right away and thus are unable to be restored to the original chemical form during the other half-reaction. Frequently, these defused products are free radicals or reactive oxygen species harmful to surrounding cells; and meanwhile, the electrodes are corroded in these irreversible Faradaic reactions.

It should be noted that stimulating electrodes work as a pair to complete the circuit. When the two electrodes are made of the same material, a half-reaction occurs on one electrode, while the other half-reaction occurs on the other electrode simultaneously during the first phase of stimulation. And the two half-reactions switch their locations during the second phase of restoration. However, the two electrodes do not need to be made of the same material and size. In such a case, two synchronized but independent reversible Faradaic reactions each take place at one electrode, while switching their own half-reactions with the phase switch of the biphasic stimulus. Accordingly, in extracellular stimulation, the neural stimulating electrode is the one placed near the neuronal target, through which the cathodic phase of the current stimulus is used to depolarize the neuron, while in intracellular stimulation, the neural stimulating electrode is placed in the neuron and depolarize the neuron in the anodic phase of the current stimulus. For more details, please see Chap. 12.

Common materials used for fabricating or functionally coating neural stimulating electrodes include platinum, iridium oxide, and conducting polymers. The associated reversible Faradaic reactions whose primary reaction products are immobilized on the electrode surface are listed below (Horch and Dhillon 2004; Guo 2017):

- Platinum:

  - Deliver negative ionic charges into the solution using a cathodic-first, charge-balanced biphasic stimulus:

$$Pt + H_2O + e^- \rightleftharpoons Pt - H + OH^-$$

    The forward direction is the cathodic phase reduction half-reaction (H-atom plating) for delivering negative ionic charges (OH^-) into the solution; and the backward direction is the anodic phase oxidation half-reaction for restoring the original reactants. During one full stimulating cycle of the two phases, the electrode and solution's constitutions and the pH are not changed.

  - Deliver positive ionic charges into the solution using an anodic-first, charge-balanced biphasic stimulus:

$$Pt + H_2O \rightleftharpoons PtO + 2H^+ + 2e^-$$

    The forward direction is the anodic phase oxidation half-reaction for delivering positive ionic charges (H^+) into the solution; and the backward direction is the cathodic phase reduction half-reaction for restoring the original reactants. During one full stimulating cycle of the two phases, the electrode and solution's constitutions and the pH are not changed.

- Iridium and iridium oxides (valence change oxides):

  - Deliver positive ionic charges into the solution using an anodic-first, charge-balanced biphasic stimulus:

$$Ir + H_2O \rightleftharpoons IrO + 2H^+ + 2e^-$$

$$IrO + H_2O \rightleftharpoons IrO_2 + 2H^+ + 2e^-$$

    The forward directions are the sequential anodic phase oxidation half-reactions for delivering positive ionic charges (H^+) into the solution; and the backward directions are the sequential cathodic phase reduction half-reactions for restoring the original reactants. During one full stimulating cycle of the two phases, the electrode and solution's constitutions and the pH are not changed.

  - Deliver negative ionic charges (OH^-) into the solution using a cathodic-first, charge-balanced biphasic stimulus:

$$IrO_2 + H_2O + 2e^- \rightleftharpoons IrO + 2OH^-$$

If an iridium electrode needs to deliver negative ionic charges into the solution using a cathodic-first, charge-balanced biphasic stimulus, it needs to be first functionalized by creating an oxide layer, as is the common practice.

- Conducting polymer PPy:

  - Deliver negative ionic charges into the solution using a cathodic-first, charge-balanced biphasic stimulus:

$$-[Py]_n^+ - + e^- \rightleftharpoons -[Py]_n -$$

The forward direction is the reduction half-reaction where the positively charged PPy backbone is reduced/neutralized causing either the doped anions in the polymer matrix to diffuse out of or the cations from the solution to diffuse into the polymer matrix to produce a negative outward current. The backward direction is the oxidation half-reaction where the neutral PPy backbone is re-oxidized causing either the anions from the solution to diffuse back into or the absorbed cations to diffuse out of the polymer matrix to produce a positive outward current. Because, within a certain voltage or current density limit, no Faradaic reactions actually take place at the polymer–electrolyte interface, this property is termed as "pseudo-capacitance". When the voltage or current density limit is exceeded, irreversible Faradaic reactions, however, can take place. A special precaution should be taken not to overdrive the PPy backbone during the anodic phase of an anodic-first, charge-balanced biphasic stimulus for delivering positive ionic charges into the solution, as the PPy backbone is prone to overoxidation.

Common unwanted irreversible Faradaic reactions occurring on the electrode surface include corrosion to electrodes, electrolysis of water, and oxidation of Cl^- (Horch and Dhillon 2004). Their reaction products readily diffuse away, and the reactions thus cannot be reversed. In these cases, the chemical composition of the biological environment is altered, creating toxic reactive products and/or causing shifts in the pH.

- Corrosion to electrodes:

$$Pt + 4Cl^- = [PtCl_4]^{2-} + 2e^-$$

- Electrolysis of water:

$$2H_2O + 2e^- = H_2 + 2OH^-$$

$$2H_2O = O_2 + 4H^+ + 4e^-$$

- Oxidation of Cl^-:

$$2Cl^- = Cl_2 + 2e^-$$

$$Cl^- + H_2O = ClO^- + 2H^+ + 2e^-$$

4.2   Electrolytic Cell for Neural Stimulation

The above reversible and irreversible Faradaic reactions all take place in the electrode's EDL, i.e. across the R_f, as a result of externally supplied electronic current. Thus, the electrode configuration for neural stimulation forms an electrolytic cell, in which the stimulating electrode functions as the working electrode, and the other electrode as the counter electrode. A reference or *GND* electrode is not necessary unless the potentials in the stimulated volume conductor need to be recorded. A two-electrode stimulating configuration, called bipolar stimulation, is very common. Additionally, the counter electrode does not need to be geometrically and materially the same as the working electrode but must also be able to conduct reversible Faradaic currents, though the specific redox reactions do not need to be the same as those occurring on the working electrode. However, the counter electrode is commonly made larger than the working electrode to reduce the current density on its surface below the neural activation threshold, as the tissue around it usually is not a target of stimulation and can be otherwise stimulated in the restoration phase of the stimulus or subjected to charge-induced injury if the current density remains high. The stimulating electrode is made small for better spatial selectivity and to boost the charge density for more efficient neural activation (see Sect. 12.3.4 in Chap. 12).

Besides the types of reversible Faradaic reactions occurring in the EDLs of both the working and counter electrodes, the electric potential across each EDL, i.e. across the R_f, is also of great concern, because it determines whether other unwanted irreversible Faradaic reactions as shown in Sect. 4.1 could happen in the EDL in the meantime. This potential is named as the *overpotential* $\eta = I_f R_f$. Usually, η is restrained within the water electrolysis window for Pt electrodes: -0.6 to 0.8 V.

It should be noted that a neural stimulating electrode should not operate under a constant DC current to protect integrity of the electrode, as a charge-balanced reverse current phase with the reverse half-reaction is required to restore the electrode and solution's compositions. Occasionally, a monophasic pulsatile stimulus is used in short-term studies, but this type of stimulus is unsuitable for long-term or clinical applications, as the stimulation deteriorates the electrode and generates cytotoxic chemical species in the surrounding solution (because the reverse half-reaction cannot be involved for restoration). Therefore, the electrolytic cell for neural stimulation should not be operated under DC conditions, either constant or pulsatile, in contrast to the general electrolytic cell such as for electroplating.

4.3 The Complete Neural Stimulating Circuit

The neural stimulating circuit is shown in Fig. 4.1. The stimulating current $I_{stim}(s)$ is a fraction of the output current $I_{out}(s)$ of the electronic stimulator, according to

$$I_{stim}(s) = \frac{Z_{out}}{Z_e + Z_x + Z_{out}} I_{out}(s) \tag{4.1}$$

where Z_e is the effective *electrode stimulating impedance*. The output voltage $V_{out}(s)$ of the stimulator is not the stimulating voltage. $V_x(s)$ is the *in situ stimulating voltage* generated by $I_{stim}(s)$ at the outer edge of the EDL of the stimulating electrode in presence of the cellular or tissue load. From the circuit, we have

$$V_x(s) = V_{out}(s) - \eta = I_{stim}(s)Z_x \tag{4.2}$$

Equation (4.2) indicates that current stimulation is preferred to voltage stimulation for the purpose of precise parameter control, as the in situ impedance Z_x in the target environment is usually unknown and may vary over the time course of stimulation. With a fixed output stimulating current $I_{stim}(s)$, the output voltage $V_{out}(s)$ and the instantanous power $p_{out}(t) = v_{out}(t)i_{out}(t)$ of the stimulator thus depend on both the overpotential η and the in situ impedance Z_x. To conserve power, η (thus the electrode stimulating impedance Z_e) needs to be minimized. Additionally, a higher η causes more risks on unwanted irreversible Faradaic reactions. Thus, these expectations pose particular requirements on the electrochemical characteristics of the stimulating electode's material, i.e. to minimize Z_e. It should be noted that to properly measure the Z_e, the testing current should generate the same level of Faradaic reactions in the electrode's EDL to involve the R_f as during the actual stimulation, as opposed to using a small testing current in the case of recording electrodes which does not involve the R_f.

In Chap. 12, we will learn that the stimulating voltage $V_x(s)$ is the actual effector on neural stimulation and how to dramatically increase the *in situ* impedance Z_x by creating a tigher electrode-neuron seal to enhance $V_x(s)$ under a fixed $I_{stim}(s)$, as an unnecessarily higher $I_{stim}(s)$ should be avoided to both conserve power and minimize risks of unwanted irreversible Faradaic reactions.

Fig. 4.1 The complete electrical stimulating circuit

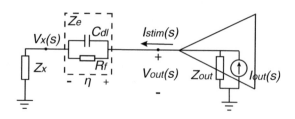

Also, for a better power efficiency, the output impedance Z_{out} of the current source is suggested to be at least one order of magnitude larger than the series impedance of $Z_e + Z_x$, according to Eq. (4.1).

4.4 Charge Injection Capacity

4.4.1 Cyclic Voltammogram (CV), CSC, and CIC

The maximal CSC of a certain electrode material is usually characterized by cyclic voltammetry (Fig. 4.2). The setup of a three-electrode electrochemical cell (Fig. 4.2a) is similar to that for EIS, with the testing electrode as the working electrode. The counter electrode is usually a large Pt mesh electrode. The reference

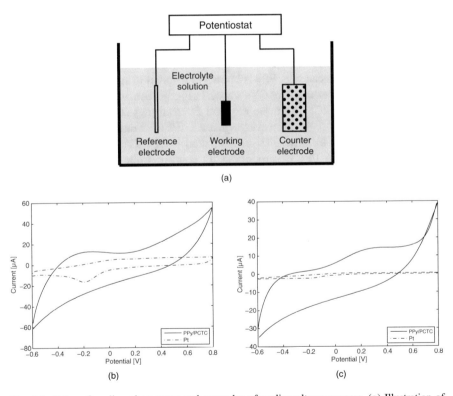

Fig. 4.2 Setup of cyclic voltammetry and examples of cyclic voltammograms. (**a**) Illustration of the basic setup. (**b**) CVs of one PPy disc electrode (1 mm diameter) and one Pt disc electrode of the same size at a voltage scan rate of 50 mV/s within the water electrolysis window. The cCSC of the PPy electrode is -48.8 mC/cm^2; and that of the Pt electrode is -5.0 mC/cm^2. (**c**) CVs at a scan rate of 1 V/s. The cathodic charge transfers are: PPy, -3.9 mC/cm^2; Pt, -1.1 mC/cm^2. (**b**) and (**c**) reproduced with permission from (Guo et al. 2014). Copyright © 2013 WILEY-VCH Verlag GmbH & Co. KGaA, Weinheim

electrode using a standard hydrogen or Ag/AgCl electrode sets the zero potential in the electrochemical cell. During the cyclic voltammetry, the potential on the working electrode is linearly increased from a starting potential up to a reverse potential and then decreased to the starting potential with the same slew rate. For neural stimulating electrodes, the potential range is set as the water electrolysis window of Pt electrodes: -0.6 to 0.8 V. The current flow through the working electrode is plotted *vs.* the ramp voltage, which is called a CV. Example CVs are shown in Fig. 4.2b, c. Such a cycle is repeated for several times to observe the evolution of the CVs. The first one or two CVs are usually discarded, as the electrode may not stabilize to its fully functional state. The area under a CV is a measure for the maximum charge delivery capacity, including both positive and negative charges. For extracellular neural stimulation, the cathodic CSC (cCSC, i.e. the total negative charges delivered in a CV) is of particular interest. This cCSC can be calculated from the CV by counting the area enclosed by the $y = 0$ line and the CV below.

The voltage scan rate for measuring the maximal CSC is usually chosen as 50 mV/s, which is a slow rate to allow complete Faradaic reactions to happen at each voltage step. In the meantime, there is another term called CIC which is a more practical electrode property for neuronal stimulation, as fast, short current pulses with a width of hundreds of microseconds are often used. This property is measured using a faster voltage scan rate. PPy is known to respond to current/voltage changes relatively slow. In the example of CVs shown in Fig. 4.2b, c, when the slow 50 mV/s scan rate is used, the cCSCs for the PPy and Pt electrodes are -48.8 mC/cm^2 and -5.0 mC/cm^2, respectively. However, when a 1 V/s scan rate is used, the cathodic CICs (cCICs) reduce to only -3.9 mC/cm^2 and -1.1 mC/cm^2, respectively. The larger decrease of the PPy electrode's cCIC conforms that this type of material responds to voltage changes slower than Pt.

4.4.2 Methods to Functionalize a Stimulating Electrode

The goal of co-designing neural stimulating electrodes and current stimulus waveforms is to make use of the reversible Faradaic reactions for a sufficient charge transfer while in the meantime avoiding the harmful irreversible Faradaic reactions. The electrode design is through a careful selection of the electrode materials and size to meet the charge delivery demand. As many neural electrodes are microfabricated nowadays and certain electrode materials may not be compatible with the microfabrication processes, the electrodes are thus commonly fabricated using a conventional conductor such as gold first and then functionalized by coating with the desired material of a high cCSC, such as platinum black, IrO$_x$, conducting polymers, or carbon nanotubes. The purpose is to enhance both the capacitive and reversible Faradic charge transfer capacities while reducing the electrode impedance. However, such a coating approach suffers from peeling and/or rubbing off issues, limiting the chronically implanted application of the electrodes.

Once the electrode material and size are determined, the stimulus waveform design is to constrain the overpotential amplitudes within the water electrolysis window and properly restore the first half-reaction using a biphasic or even multiphasic waveform design while considering the kinetics of the half-reaction pair. Cyclic voltammetry using the sample electrodes can help to obtain critical information to inform this design process.

4.5 Summary

- The different requirements for recording and stimulating electrodes stem from the different amount of current needed for conversion.
- Neural stimulating electrodes employ reversible Faradaic reactions with charge-balanced biphasic current stimulus waveforms to inject currents at a substantial current density primarily through the R_f path in the electrode's EDL.
- Stimulating electrodes work as a pair to complete the circuit. The two electrodes do not need to be made of the same material and size, but the two synchronized yet independent reversible Faradaic reactions each taking place at one electrode need to produce a complementary overall current. In extracellular stimulation, the neural stimulating electrode needs to be placed near the neural target, through which the cathodic phase of the current stimulus is used to depolarize the neuron. The counter electrode is commonly made larger than the working electrode to reduce the current density on its surface below the neural activation threshold, as the tissue around it usually is not a target of stimulation and can be otherwise stimulated in the restoration phase of the stimulus or subjected to charge-induced injury if the current density remains high. The stimulating electrode is made small for better spatial selectivity and to boost the charge density for more efficient neural activation.
- The electrode configuration for neural stimulation forms an electrolytic cell, which should not be operated under DC conditions, either constant or pulsatile.
- The stimulating voltage $V_x(s)$ is the actual effector on neural stimulation and is generated by $I_{stim}(s)$ at the outer edge of the EDL of the stimulating electrode.
- For precise parameter control, current stimulation is preferred to voltage stimulation, as the *in situ* impedance Z_x in the target environment is usually unknown and may vary over the time course of stimulation.
- To conserve power and minimize unwanted irreversible Faradaic reactions in the EDL, the overpotential η across the electrode's EDL (thus the electrode impedance Z_e) needs to be minimized.
- The maximal CSC of a certain electrode material is usually characterized by cyclic voltammetry. For extracellular neural stimulation, the cCSC is of particular interest. This cCSC can be calculated from the CV by counting the area enclosed by the $y = 0$ line and the CV below.

- CIC is a more practical electrode property for neuronal stimulation, as fast, short current pulses with a width of hundreds of microseconds are often used. This property is measured using a faster voltage scan rate, e.g. 1 V/s.
- Co-design of neural stimulating electrodes and current stimulus waveforms is needed to make use of the reversible Faradaic reactions for a sufficient charge transfer while in the meantime avoiding the harmful irreversible Faradaic reactions.
- Stimulating electrodes are often functionalized after fabrication using a high cCSC material, such as platinum black, IrO_x, conducting polymers, or carbon nanotubes. The purpose is to enhance both the capacitive and reversible Faradic charge transfer capacities while reducing the electrode impedance.

Exercises

4.1 For neural stimulating electrodes, explain:

    (a) The charge transfer mechanisms in the EDL at the electrode–electrolyte interface.

    (b) Why is a high CIC desired?

    (c) How to improve the CIC?

    (d) Why are platinum and iridium oxides preferred as the electrode materials?

4.2 For neural stimulating electrodes,

    (a) What are the differences between stimulating and recording electrodes?

    (b) Can an electrode be used for both stimulating and recording?

    (c) How much is the capacitive CSC of a smooth metal electrode? What mechanisms are employed to meet the stimulating charge density requirement?

    (d) What effects should be avoided? When these effects happen, what are the consequences?

    (e) How is the stimulating safety criteria set?

4.3 The reversible electrochemical redox reactions associated with iridium and its oxides in a physiological solution are more complicated than what have been listed in Sect. 4.1. Please search on the internet and dig into relevant literature to delineate a more complete reaction picture.

4.4 Considering the neural stimulating circuit in Fig. 4.1,

    (a) Explain why the electrode impedance Z_e needs to be minimized.

    (b) Consider the electrode's EDL, how to minimize the Z_e?

    (c) What is the goal that the stimulating circuit wants to achieve?

4.5 Explain:

    (a) Why are current-controlled electrical stimuli preferred to voltage-controlled ones?

(b) What aspects need to consider when designing current stimulus wave-forms? [Hint: The lower of cCSC and anodic CSC (aCSC) constrains the stimulus waveform design.]

(c) Why are charge density and charge per phase responsible for electrical stimulation induced neural injury?

4.6 How is the stimulating capability of an electrode characterized? Specifically for neural stimulation, what physical quantity is used and how is it calculated?

4.7 For an MEA consisting of multiple smooth metal microelectrodes designed for neural stimulation, e.g., the NeuroNexus Probes, explain:

(a) Why does the electrochemical impedance of a stimulating electrode need to be minimized?

(b) What aspects need to consider in order to improve the performances of the MEA? What are the methods used to achieve such improvements?

(c) If the MEA is to be used for simultaneous neural stimulation and recording (i.e., some of the electrodes are used for stimulation, while the others are used for recording the elicited neural responses at the same time), what is (are) the potential problem(s)?

4.8 Given that

1. The cCSC of a smooth Au electrode is $20\mu C/cm^2$;
2. The cCSC of a Pt electrode is $5.0\ mC/cm^2$;
3. The cCSC of an iridium oxide electrode is $28.8\ mC/cm^2$;
4. The cCSC of a PEDOT electrode is $75.6\ mC/cm^2$;
5. The charge threshold required to activate an innervated muscle is $20\mu C$; and
6. The charge threshold required to activate the same but denervated muscle is $600\mu C$.

Consider a disc electrode placed on the surface of the muscle for epimysial electrical stimulation, answer the following questions:

(a) Calculate the minimum electrode areas needed for activating the inner-vated muscle using electrodes made of Au, Pt, iridium oxide, and PEDOT, respectively.

(b) Calculate the minimum electrode areas needed for activating the dener-vated muscle using electrodes made of Au, Pt, iridium oxide, and PEDOT, respectively.

(c) What conclusion can you draw with respect to the selection of electrode (coating) materials?

(d) With the electrode size fixed, to deliver the same amount of negative charges, what is the advantage of using a material of a higher cCSC?

4.9 Consider a cochlear implant electrode as shown in Fig. 1.2a. The electrode material is smooth Pt. According to Fig. 4.2c, the cCIC of Pt is -1.1 mC/cm². Work on the following design problems:

(a) Search on the internet and literature to find the electrode's dimensions and calculate the electrode's surface area and its cCIC.

(b) Find the amount of negative charges needed to extracellularly activate an auditory nerve fiber by the electrode. Can this electrode deliver this amount of charge? What reversible Faradaic reaction is employed?

(c) Design a charge-balanced, biphasic current stimulus waveform to achieve the stimulation goal, and state which phase of the current stimulus recruits which half-reaction.

(d) To deliver the required amount of negative charges, numerous combinations of the cathodic current amplitude and phase length can be used. What precaution should be taken in making the choice then? [Hint: consider the current density and overpotential.]

(e) Consider the overall design of the cochlear implant, what aspects could be optimized in order to lower the amount of negative charges needed to extracellularly activate an auditory nerve fiber by the electrode? When the required charge amount is reduced, what are the benefits?

References

Guo L (2017) Conducting Polymers as Smart Materials for Tissue Engineering. In: Wang Q (ed) Fundamental Principles of Smart Materials for Tissue Engineering vol RSC Smart Materials Series. Royal Society of Chemistry,

Guo L, Ma M, Zhang N, Langer R, Anderson DG (2014) Stretchable polymeric multielectrode array for conformal neural interfacing. Advanced materials 26 (9):1427-1433. doi:https://doi.org/10.1002/adma.201304140

Horch KW, Dhillon GS (2004) Neuroprosthetics: theory and practice. World Scientific.

Part II
Principles of Electrical Neural Recording

Chapter 5
Intracellular Recording

A neuron's signature transmembrane potentials, including the DC resting membrane potential and the AC AP, can be unambiguously measured using an electrode that directly accesses to the cytosol. Although this technique has been well established, pitfalls exist for beginners and students to develop an accurate and complete understanding on the recording mechanisms, particularly considering the variations of recordings caused by limitations of the tool and the user's inappropriate configuration. In this chapter, we analyze the different recording setups based on models of the complete recording circuits, including the equivalent electrical circuits of the neuron and electrodes. Such an approach gives us the capability to derive analytical solutions to the recorded signals and thus elucidate the recording mechanism and the relationship of the recorded signals to the transmembrane voltage. This knowledge further helps us to understand proper configuration of the recording systems and identify the system parameters critical to the recording result.

5.1 DC Recording: The Resting Membrane Potential

Figure 5.1 depicts the configuration for intracellularly recording the resting transmembrane potential with a complete equivalent circuit model. As described in Chap. 2 Sect. 2.2, the neuronal membrane for DC analysis is simply modeled as a capacitor C_m in parallel with a serial connection of a resistor R_m and an equivalent Nernst potential E_m (Fig. 2.2). The immediate extracellular space of the neuron is grounded. Upon breaking the cell membrane by the micropipette (i.e., the patch-clamp electrode) through a negative pressure suction, the configuration nonetheless slightly attenuates the original resting membrane potential V_{mrest} due to two leaking paths including the seal around the micropipette and the micropipette lumen. For the best result to avoid substantial current leakages to attenuate the resting membrane potential, the micropipette should be configured as a whole-cell patch clamp with (1) the seal resistance R_{seal} in the GΩ range and (2) the lumen solution resistance R_{fill}

© The Author(s), under exclusive license to Springer Nature Switzerland AG 2022
L. Guo, *Principles of Electrical Neural Interfacing*,
https://doi.org/10.1007/978-3-030-77677-0_5

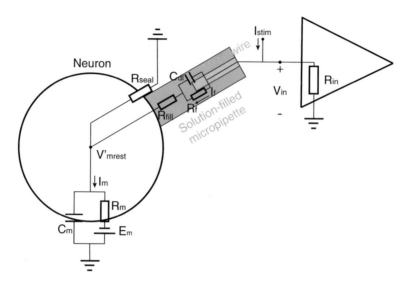

Fig. 5.1 Whole-cell patch-clamp configuration under current clamp mode for recording the resting membrane potential

greater than 10 MΩ. The high resistance R_{fill} in the micropipette tip also minimizes the filled solution to contaminate and disturb the composition of the cytosol through diffusion.

Because the net transmembrane current I_m is zero at rest or equilibrium, the recording electrode with access to the cytosol cannot receive any ionic current through its EDL capacitor C_{dl} after stabilizing into the steady state; and thus the membrane voltage V'_{mrest} is simply imposed across the C_{dl}, resulting in a zero potential reading across R_{in} at the input terminal of the amplifier. To circumvent this problem, a small DC current I_{stim} is injected through the recording electrode into the cell, activating the Faradaic resistance in the electrode's EDL. This configuration is called the *current clamp* mode or the *Faradaic regime* (Robinson et al. 2012), as the transmembrane current I_m is "clamped" to the injected DC current I_f (note, the current leakage through the GΩ R_{seal} is ignored):

$$I_m \approx I_f = I_{stim} - \frac{V_{in}}{R_{in}} \tag{5.1}$$

Apparently, the resting membrane potential V_{mrest} is further affected by this injected current I_f, and we denote the new equilibrium potential as V'_{mrest}. To minimize the effect of I_f on V_{mrest}, I_f should be set to the lowest level that is sufficient to activate R_f. Therefore, this current method only provides an estimation of the resting membrane potential. Other more accurate methods for measuring the resting membrane potential include perforated patch clamp (Akaike 1996; Akaike and Harata 1994) and using single channel currents in cell-attached mode (Tyzio et al. 2008).

The DC current I_{stim} is applied as a rectangular current pulse (it is also a current step with a defined length). At the beginning and end of the step, there is a transient response in both the neuron and the electrode recording circuit, respectively. We should wait for the systems to settle down in their steady state before performing the measurement.

In Fig. 5.1, the stabilized membrane potential V'_{mrest} is determined as

$$V'_{mrest} = V_{in} - \left(R_f + R_{fill}\right)I_f = \frac{R_f + R_{fill} + R_{in}}{R_{in}} V_{in} - \left(R_f + R_{fill}\right)I_{stim} \quad (5.2)$$

This requires V_{in}, I_{stim}, R_{in}, R_f, and R_{fill} to be known or measured.

With the current method, in order to obtain a more accurate measurement of the intracellular potential V_{mrest}, i.e. making V'_{mrest} a close approximation to V_{mrest}, three precautions need to be taken: (1) as I_{stim} depolarizes the V_{mrest}, its amplitude and duration need to be made small—only sufficient to activate the R_f and settle down in the steady state for making the measurement. (2) The seal resistance R_{seal} needs to be made large enough to prevent substantial current leakage around the glass micropipette, preferably in the GΩ range. And (3) the amplifier's input resistance R_{in} needs to be at least comparable to, if not much larger than, the combined resistance $R_f + R_{fill} + (R_m \parallel R_{seal})$, in order to drain the major fraction of I_{stim} through this combined resistance to minimize I_{stim}'s effect on the amplifier's input. Also, note that R_f needs to be small to minimize the overpotential η.

According to its role in this method, this type of amplifier is different from those used for AC recordings below and is named as *DC amplifier*. It often has little or no amplification to the voltage input and is frequently setup as a unity gain amplifier or voltage follower to directly read out the V_{in}.

5.2   AC Recording

The dynamic transmembrane voltage change $v_m(t)$ (its frequency-domain representation, i.e. its Laplace transform, is $V_m(s)$) is of particular interest, as it reflects the process of neuronal information processing. $V_m(s)$ includes two phases where the membrane is modeled differently (see Chap. 2, Sect. 2.3): $V_{msub}(s)$ and $V_{mAP}(s)$ to account for the subthreshold and suprathreshold transmembrane voltage changes, respectively. Accordingly, we will perform our analyses on these two phases separately.

5.2.1   How the AC $V_m(s)$ Is Generated

Before we look into the intracellular recording mechanisms, we need to find out how the intracellular transmembrane voltage $V_m(s)$ is generated in a neuron, in the absence of an inserted recording microelectrode.

Referring to Fig. 5.2a, during subthreshold membrane voltage changes, a current $I_{stim}(s)$ is injected into the neuron, either externally or through synaptic inputs. This current flows across the membrane to produce the AC transmembrane voltage so that

$$V_{msub}(s) = \left(\frac{1}{sC_m} \parallel R_m \right) I_{stim}(s) = \frac{R_m}{1 + sC_m R_m} I_{stim}(s) \tag{5.3}$$

In the range of neuronal characteristic frequencies around 1 kHz, $\frac{1}{sC_m}$ is one order of magnitude smaller than R_m (Guo 2020) so that

$$V_{msub}(s) \approx \frac{1}{sC_m} I_{stim}(s) \tag{5.4}$$

Referring to Fig. 5.2b, during AP, the ionic current entering the neuron is $I_{AP}(s) = I_{Na}(s) - I_K(s) - I_{Cl}(s)$ (see Sect. 2.3.2 in Chap. 2) which charges the C_m to produce the AC transmembrane voltage change so that

$$V_{mAP}(s) = \frac{1}{sC_m} I_{AP}(s) \tag{5.5}$$

Fig. 5.2 Equivalent electrical circuits of a neuron for deriving the AC transmembrane voltages. (**a**) Subthreshold phase. (**b**) AP phase

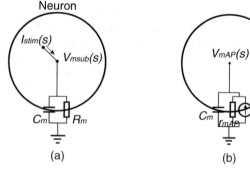

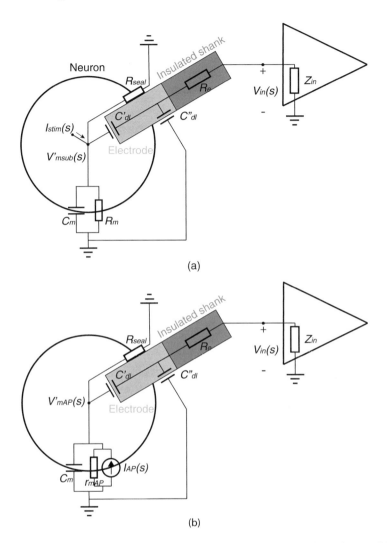

Fig. 5.3 Configurations and equivalent circuits of intracellular AC recording using a solid-state microwire electrode. (**a**) Subthreshold phase. (**b**) AP phase

5.2.2   Intracellular Recording Using a Solid-State Microwire Electrode

Figure 5.3 illustrates the configurations of intracellular AC recording directly using a solid-state microelectrode such as a microwire electrode during subthreshold and AP phases, respectively. Although intracellular AC recording seems to be a straightforward process, where the internal potential of a neuron is sensed via a voltage-divider circuit formed by the electrode and amplifier-input impedances, it should be noted that this overly simplified impression is actually an approximation to the more complicated scenarios illustrated in Fig. 5.3a, b, where four important aspects are brought up. First, the subthreshold and AP phases need to be analyzed separately, as the plasma membrane's electrical properties are different in these two phases. Second, a seal resistance R_{seal} is used to model the ionic leakage around the electrode shank. Third, with introduction of the electrode recording system and the associated ionic leakage through R_{seal}, the internal membrane potential $V_m(s)$ is slightly lowered to $V'_m(s)$ in each phase. Fourth, a portion of the electrode may be left in the extracellular space, as shown in Fig. 5.3. Thus, we analyze these configurations in detail below.

In Fig. 5.3a for the subthreshold depolarization phase, a current $I_{stim}(s)$ is injected into the cell, from either an external stimulator or synaptic inputs, to charge the inner side of the plasma membrane. In absence of the inserted electrode, Eq. (5.3) holds. With the electrode inserted, let us first assume that the electrode is entirely in the cell, so that $C''_{dl} = 0$. The internal potential becomes

$$
\begin{aligned}
V'_{msub}(s) &= \left[\frac{1}{sC_m} \parallel R_m \parallel R_{seal} \parallel (Z_e + Z_{in}) \right] I_{stim}(s) \\
&= \frac{R_m R_{seal}(Z_e + Z_{in})}{R_m R_{seal} + (Z_e + Z_{in})(R_m + (1 + sC_m R_m)R_{seal})} I_{stim}(s) \\
&= \frac{R_{seal}(Z_e + Z_{in})(1 + sC_m R_m)}{R_m R_{seal} + (Z_e + Z_{in})(R_m + (1 + sC_m R_m)R_{seal})} \frac{R_m}{1 + sC_m R_m} I_{stim}(s) \\
&= \frac{1}{\frac{R_{seal} + Z_e + Z_{in}}{\left(\frac{1}{R_m} + sC_m\right)R_{seal}(Z_e + Z_{in})} + 1} V_{msub}(s) \approx \frac{1}{\frac{R_{seal} + Z_e + Z_{in}}{sC_m R_{seal}(Z_e + Z_{in})} + 1} V_{msub}(s)
\end{aligned}
$$

$$(5.6)$$

where $Z_e = \frac{1}{sC'_{dl}} + R_e$ is the electrode's recording impedance. Thus, $V'_{msub}(s)$ is attenuated to a fraction of $V_{msub}(s)$.

Referring to Fig. 5.3b for the AP phase, in absence of the inserted electrode, Eq. (5.5) holds. With the electrode inserted, let us also assume that the electrode is entirely in the cell, so that $C''_{dl} = 0$. The internal potential becomes

$$V'_{mAP}(s) = \left[\frac{1}{sC_m} \parallel R_{seal} \parallel (Z_e + Z_{in})\right] I_{AP}(s)$$

$$= \frac{\frac{R_{seal}}{1+sC_mR_{seal}}(Z_e + Z_{in})}{\frac{R_{seal}}{1+sC_mR_{seal}} + (Z_e + Z_{in})} I_{AP}(s)$$

$$= \frac{sC_mR_{seal}(Z_e + Z_{in})}{R_{seal} + (1 + sC_mR_{seal})(Z_e + Z_{in})} \frac{1}{sC_m} I_{AP}(s)$$

$$= \frac{1}{\frac{R_{seal}+Z_e+Z_{in}}{sC_mR_{seal}(Z_e+Z_{in})} + 1} V_{mAP}(s) \qquad (5.7)$$

Thus, $V'_{mAP}(s)$ is also a fraction of $V_{mAP}(s)$.

Both Eqs. (5.6) and (5.7) have the same fraction factor, where the values of R_{seal} and $Z_e + Z_{in}$ have significant impacts to this fraction. Only when $\left|\frac{R_{seal}+Z_e+Z_{in}}{sC_mR_{seal}(Z_e+Z_{in})}\right| \ll 1$, $V'_{msub}(s) \approx V_{msub}(s)$ and $V'_{mAP}(s) \approx V_{mAP}(s)$. Usually, $Z_e + Z_{in}$ is very large, thus the leakage through the electrode path is minimum. If R_{seal} is in the GΩ scale, such as in the configuration of whole-cell patch-clamp recording, the leakage through R_{seal} is also neglectable. However, in other recording configurations, such as using a tungsten microneedle electrode or a glass micropipette electrode, the membrane seal around the electrode shank can be poor (e.g., an R_{seal} in the 100 MΩ scale or lower), leading to a much attenuated $V'_{msub}(s)$ and $V'_{mAP}(s)$.

The recorded signals are

$$V_{insub}(s) = \frac{Z_{in}}{Z_e + Z_{in}} V'_{msub}(s) \approx \frac{Z_{in}}{\frac{R_{seal}+Z_e+Z_{in}}{sC_mR_{seal}} + Z_e + Z_{in}} V_{msub}(s) \qquad (5.8)$$

$$V_{inAP}(s) = \frac{Z_{in}}{Z_e + Z_{in}} V'_{mAP}(s) = \frac{Z_{in}}{\frac{R_{seal}+Z_e+Z_{in}}{sC_mR_{seal}} + Z_e + Z_{in}} V_{mAP}(s) \qquad (5.9)$$

If the electrode is only partially inserted into the cell, $C''_{dl} \neq 0$. If the immediate extracellular space is grounded, C''_{dl} will act to shunt the amplifier's input impedance Z_{in}. If the ground is place relatively farther away from the cell, C''_{dl} will pick up an eAP that is built up across the spreading resistance R_s of the extracellular solution. However, in an open extracellular space such as in a Petri dish, this eAP component cannot be detected above the noise floor of the recording circuit (see Exercise 6.5 in Chap. 6). Thus, overall, C''_{dl} has a negative shunting effect to the amplifier's input impedance.

Last, in this type of AC recording, the input impedance Z_{in} of the differential amplifier needs to be very large, e.g., >10 MΩ at 1 kHz. The role of this type of amplifier is quite different from that of the DC amplifier in Sect. 5.1, and it is named as *AC amplifier*.

5.2.3   Intracellular Recording Using Whole-Cell Patch-Clamp and Glass Micropipettes

Figure 5.4 illustrates the configurations of intracellular AC recording using whole-cell patch-clamp and glass micropipettes, where a solution-filled glass micropipette is used. There is an additional solution resistance R_{fill} in serial connection before C_{dl} to represent the filled solution in the fine glass micropipette. Nonetheless, this R_{fill} can be lumped into R_e for simplicity. A prominent difference between whole-cell patch-clamp and glass micropipettes is the magnitude of the seal resistance R_{seal}

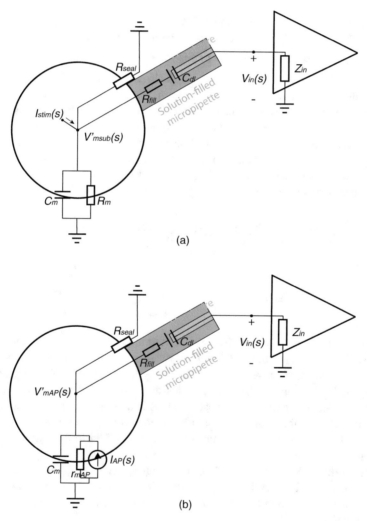

Fig. 5.4 Configurations and equivalent circuits of intracellular AC recording using whole-cell patch-clamp and glass micropipettes. (**a**) Subthreshold phase. (**b**) AP phase

modeling the ionic leakage around the micropipette's shank. The advantage of whole-cell patch-clamp recording over other recording approaches, including glass micropipette recording, is the much higher $G\Omega$ R_{seal}. Thus, the intracellular potential is better preserved from attenuation through leakage. Furthermore, recording using a solution-filled micropipette has two additional advantages: (1) the possible shunting electrode capacitance C''_{dl} in Fig. 5.3 is avoided; and (2) as the metal wire electrode inserted in the latter section of the glass pipette can be large in dimensions, C_{dl} is much larger comparing to the C'_{dl} of a solid-state microelectrode in Fig. 5.3, leading to a lower overall electrode recording impedance Z_e that can improve the recording quality. Equations (5.6)–(5.9) all apply to Fig. 5.4 with the substitution of $Z_e = R_{fill} + \frac{1}{sC_{dl}}$.

5.3 Summary

- Whole-cell patch-clamp configuration under current clamp mode is used for recording the resting membrane potential. A small DC current I_{stim} is injected through the recording electrode into the cell to activate the Faradaic resistance in the electrode's EDL, which is essential for the measurement.
- The type of amplifier used in measuring the resting membrane potential is different from those used for AC recordings and is named as DC amplifier. It is frequently set up as a unity gain amplifier or voltage follower to directly read out the V_{in}.
- During subthreshold membrane voltage changes, $V_{msub}(s) \approx \frac{1}{sC_m} I_{stim}(s)$, while during AP, $V_{mAP}(s) = \frac{1}{sC_m} I_{AP}(s)$.
- After a microelectrode's impalement of the membrane, both $V'_{msub}(s)$ and $V'_{mAP}(s)$ are attenuated to a fraction of $V_{msub}(s)$ and $V_{mAP}(s)$, respectively. The values of R_{seal} and $Z_e + Z_{in}$ have significant impacts to this fraction; and both need to be made large.
- If the electrode is only partially inserted into the cell, $C''_{dl} \neq 0$. C''_{dl} has a negative shunting effect to the amplifier's input impedance.
- The prominent advantage of whole-cell patch-clamp recording over other recording approaches, including glass micropipette recording, is the much higher $G\Omega$ R_{seal}. Furthermore, recording using a solution-filled micropipette has two additional advantages: (1) the possible shunting electrode capacitance C''_{dl} is avoided; and (2) as the metal wire electrode inserted in the latter section of the glass pipette can be large in dimensions, C_{dl} is much larger comparing to the C'_{dl} of a solid-state microelectrode, leading to a lower overall electrode recording impedance Z_e that can improve the recording quality.

Exercises
5.1 Design an experiment to measure the passive membrane capacitance C_m and resistance R_m of a neuron.

5.2 Considering the whole-cell patch-clamp configuration in Fig. 5.1 for recording the resting membrane potential,

    (a) What are the limitations of this approach under the current clamp mode?

    (b) Is it possible to measure and calculate the resting membrane potential without using the current clamp mode? [Hint: consider the transient response of the recording circuit at the moment when the micropipette impales the membrane.]

5.3 Describe another method for measuring the resting membrane potential. How accurate is it? What are the potential pitfalls? What parameters need to be improved in order to improve the accuracy? How?

5.4 Referring to Fig. 5.2, how is the AC $V_m(s)$ generated in the subthreshold and suprathreshold phases, respectively?

5.5 For intracellular AC recording,

    (a) Comparing to whole-cell patch-clamp recording, what are the disadvantages of recording using an impaled glass micropipette?

    (b) Comparing to whole-cell patch-clamp recording, what are the disadvantages of recording using a solid-state microwire electrode?

5.6 Referring to Fig. 5.3 on intracellular recording using a solid-state microwire electrode, what parameters need to be strictly controlled in order to obtain a high-quality recording?

References

Akaike N (1996) Gramicidin perforated patch recording and intracellular chloride activity in excitable cells. Progress in biophysics and molecular biology 65 (3):251-264

Akaike N, Harata N (1994) Nystatin perforated patch recording and its applications to analyses of intracellular mechanisms. The Japanese journal of physiology 44 (5):433-473

Guo L (2020) Perspectives on electrical neural recording: a revisit to the fundamental concepts. J Neural Eng 17 (1):013001. doi:https://doi.org/10.1088/1741-2552/ab702f

Robinson JT, Jorgolli M, Shalek AK, Yoon MH, Gertner RS, Park H (2012) Vertical nanowire electrode arrays as a scalable platform for intracellular interfacing to neuronal circuits. Nature nanotechnology 7 (3):180-184

Tyzio R, Minlebaev M, Rheims S, Ivanov A, Jorquera I, Holmes GL, Zilberter Y, Ben-Ari Y, Khazipov R (2008) Postnatal changes in somatic γ-aminobutyric acid signalling in the rat hippocampus. European Journal of Neuroscience 27 (10):2515-2528

Chapter 6
Extracellular Recording

Comparing to intracellular recording, on the one hand, extracellular recording is more prevalent in both fundamental neuroscience investigations and neural prosthetics due to easier accessibility and capability for longer period of interfacing; but on the other hand, as the recording electrodes are more diverse and the recording setups are more variant, the recording mechanisms and resulted signals are more poorly understood. For many people who use extracellular recording techniques very often or who develop new neural recording electrodes, such ambiguity and confusion persist. This knowledge gap unfortunately has led to flawed neural electrode designs and characterizations, inaccurate interpretations of the recordings, and inconsistencies in the conclusions reached by different research groups. This chapter thus sets out to address this knowledge gap to mitigate confusions and promote understandings by using a rigorous equivalent electrical circuit model to derive analytical solutions to this critical problem. The resulting analytical solutions reveal the relationships between the extracellular field potentials (eFPs) and the transmembrane voltage changes including the intracellular action potential (iAP), as well as the key parameters affecting the recording quality.

We start to find out the basic relationships between the eFPs and transmembrane voltage changes by considering an idealized spherical neuron suspended in an infinite electrolyte solution. Then, we will particularly explore the common extracellular recording configuration using a substrate-integrated planar microelectrode. Lastly, we will analyze the key parameters affecting the recording quality in the case of using a substrate-integrated planar microelectrode.

L. Guo, *Principles of Electrical Neural Interfacing*,
https://doi.org/10.1007/978-3-030-77677-0_6

6.1　Basic Relationships Between eFPs and Transmembrane Voltage Changes

As shown in Fig. 6.1, to derive the eFP $V_X(s)$ at an extracellular Point X, we need first to find the net outward transmembrane current $I_X(s)$ during the subthreshold and AP phases, respectively. Assuming the extracellular Point X is located on an isopotential sphere through which the $I_X(s)$ flows outward, the potential $V_X(s)$ on this sphere is thus determined by $I_X(s)$ and equal to $V_X(s) = R_s I_X(s)$, where R_s is the spreading resistance from this isopotential sphere to the *GND* at the infinity. In the subthreshold phase, $I_{Xsub}(s) = I_{stim}(s)$; and during the AP phase, $I_{XAP}(s) = -I_{AP}(s)$. According to $V_X(s) = R_s I_X(s)$, $I_{Xsub}(s)$ and $I_{XAP}(s)$ produce two sequential eFP phases on the isopotential sphere with

$$V_{Xsub}(s) = R_s I_{Xsub}(s) = R_s I_{stim}(s) \tag{6.1}$$

$$V_{XAP}(s) = R_s I_{XAP}(s) = -R_s I_{AP}(s) \tag{6.2}$$

According to Eq. (5.3), during the subthreshold phase, we have $I_{stim}(s) = \left(\frac{1}{R_m} + sC_m\right)V_{msub}(s)$, thus

$$V_{Xsub}(s) = R_s \left(\frac{1}{R_m} + sC_m\right)V_{msub}(s) \tag{6.3}$$

At neuronal frequencies, Eq. (6.3) approximates to

$$V_{Xsub}(s) \approx R_s C_m \cdot s V_{msub}(s) \tag{6.4}$$

According to Eq. (5.5), during the AP phase, we have $I_{AP}(s) = C_m \cdot sV_{mAP}(s)$, thus

Fig. 6.1 Equivalent circuits for deriving the basic relationships between the eFPs and transmembrane voltage changes

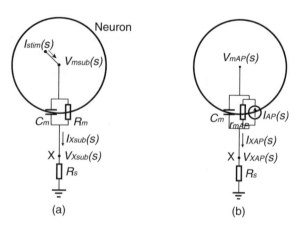

(a)　　　　　　　(b)

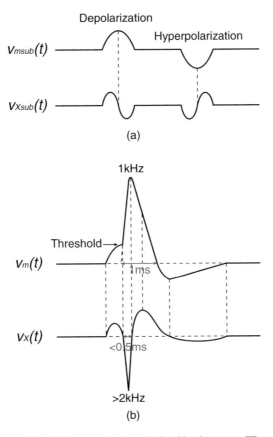

Fig. 6.2 Illustrations of the first time derivative relationships between eFPs and transmembrane voltage changes. Not to scale. (**a**) Subthreshold depolarization and hyperpolarization, e.g., EPSP and IPSP. The eFP is proportional to the first time derivative of the transmembrane voltage change. (**b**) The full time course of an AP. In the subthreshold phase, the eFP is proportional to the first time derivative of the transmembrane voltage change, whereas in the AP phase, it is proportional to the negative first time derivative of the iAP. The base width of the major iAP peak is ~1 ms, corresponding to a characteristic frequency of ~1 kHz, whereas the base width of the major eAP peak is less than 0.5 ms, corresponding to a characteristic frequency greater than 2 kHz. Note, the major eAP peak does not correspond to the major iAP peak which actually maps to the transiting zero point in the eAP

$$V_{XAP}(s) = -R_s C_m \cdot s V_{mAP}(s) \tag{6.5}$$

From Eqs. (6.4) and (6.5), we see that during the subthreshold phase, the eFP $v_{Xsub}(t)$ is proportional to the *first time derivative* (note: $sV(s) \overset{\mathcal{L}^{-1}}{\leftrightarrow} \frac{dv(t)}{dt}$) of the intracellular transmembrane voltage change $v_{msub}(t)$ (Fig. 6.2a), whereas during the AP phase, the eFP (also called eAP) $v_{XAP}(t)$ is proportional to the *negative first time derivative* of the iAP $v_{mAP}(t)$ (Fig. 6.2b). These relationships result because the same net transmembrane current produces the $v_m(t)$ by charging the inner surface of the membrane capacitor C_m, whereas producing the $v_X(t)$ through the R_s. The magnitudes of the peaks of $v_X(t)$ depend on the overall cell surface area (determines C_m), the distance that the isopotential sphere is away from the outer cell surface (determines R_s), as well as the maximum rising/falling slopes of the transmembrane voltage changes (determine $\frac{dv(t)}{dt}$) (see Exercise 6.5). Note that, according to Eq. (6.4), *any extracellularly recorded subthreshold potentials, e.g., synaptic potentials, are also biphasic in their full cycle* (Fig. 6.2a).

In Fig. 6.2b, the major peak (positive) of the iAP has a base width of ~1 ms, which corresponds to the frequently referred characteristic AP frequency of ~1 kHz. However, the major peak (negative) of the eAP has a base width less than 0.5 ms, which corresponds to a characteristic frequency greater than 2 kHz. Furthermore, *the major eAP peak does not correspond to the major iAP peak which actually maps to the transiting zero point in the eAP*. For conventional extracellular recording configurations where an ultra-tight seal is not formed between the neuronal membrane and the microelectrode, the first positive peak and the second negative peak of the eFP $v_X(t)$ in Fig. 6.2b cannot be detected above the noise level of the recording electronics; and as a result, the frequently seen eAPs only show a major negative peak (i.e., the first negative peak in Fig. 6.2b) and a minor positive peak (i.e., the second positive peak in Fig. 6.2b). Fortunately, these two missing peaks can be detected using advanced nanoprotrusion electrodes which form an ultra-tight seal with the neuronal membrane (see Chap. 9).

A frequently asked question is why the amplitude of the eAP $v_X(t)$ is so small comparing to that of the iAP $v_m(t)$. According to Eq. (6.5), this is because the coefficient $R_s C_m$ is very small and the $\frac{dv_{mAP}(t)}{dt}$ is limited (see Exercise 6.5). An alternative interpretation according to Eq. (6.2) is that (1) the net outward transmembrane current $i_{XAP}(t) = -i_{AP}(t)$ is substantially diminished due to mutual cancellation of the opposing $i_{Na}(t)$ and $i_K(t)$, and (2) R_s is small. Moreover, as $i_{XAP}(t)$ propagates into the resistive extracellular space to generate a field potential in the vicinity of the cell according to $v_{XAP}(t) = R_s i_{XAP}(t) = \frac{i_{XAP}(t)}{4\pi\sigma r}$, where σ is the conductivity of the electrolyte and r is the distance that Point X is away from the location of the equivalent (point) current source (Horch and Dhillon 2004), the amplitude of $v_{XAP}(t)$ decays with a $1/r$ characteristic. The actual decay with distance from the soma in in vitro and in vivo settings should be faster than this $1/r$ characteristic of a Coulomb potential around a point source with a current sink at the infinity (Horch and Dhillon 2004), due to the presence of adjacent current sinks (*e.g.*, other cells such as glial cells or cellular processes). The edge of this potential field is set by the ionic diffusion limit of $i_{XAP}(t)$ in the electrolyte. At this edge, $i_{XAP}(t)$ attenuates to zero, thus this edge is shorted to the *GND* at the infinity by the intermediate electrolyte solution.

It should be noted that we cannot directly compare the iAP with the eAP recorded by a free-standing point electrode in an open electrolyte, because (1) the two recording systems (Figs. 5.4 and 6.1) place their *GND*s at different locations, so that the two types of recorded potentials do not have a common potential reference to compare with and (2) the two different *GND* placements are incompatible in the same system for simultaneous recordings (i.e., if the *GND* is placed adjacent to the cell, as in the whole-cell patch-clamp recording setup in Fig. 5.4, the eFP at the Point X in Fig. 6.1 would be shorted to the *GND* due to the very low resistance of the electrolyte in-between). Fortunately, the transmembrane voltage $v_m(t)$ is a physical quantity independent from the *GND* placement, so that we can still derive a quantitative relationship between it and the eFP $v_X(t)$ thanks to the shared net transmembrane current. However, particular to the sealed recording environment of planar substrate microelectrodes (see Sect. 6.2 below), extracellular and intracellular recordings can be performed in the same setup, thanks to the moderately high electrical resistance R_{seal} (e.g., 0.1–1.2 MΩ (Dipalo et al. 2017; Hai et al. 2009)) of the cell membrane-substrate seal that separates the recording electrode surface from the exterior bulk electrolyte where the *GND* is placed as part of the intracellular recording setup.

6.2 Extracellular Recording Using a Planar Substrate Microelectrode

Modeling of the neuron-substrate electrode junctional interfaces has been reported by several groups over the past three decades (Hierlemann et al. 2011; Spira and Hai 2013; Fromherz 2002; Grattarola and Martinoia 1993; Jenkner and Fromherz 1997; Hai et al. 2010a, b). However, these mere simulation work did not provide an intuitive analytical solution that can offer a clear image on the recording mechanism, nature of the signal, and interplays between key interface parameters. Thus, this was the goal of a recent effort (Guo 2020a).

Fig. 6.3 illustrates the lumped-parameter equivalent electrical circuit models in pursuit of a closed-form analytical solution. During subthreshold depolarization (Fig. 6.3a), a positive ionic current $I_{stim}(s)$, either applied artificially or from synaptic inputs, is injected intracellularly to depolarize the membrane. Consequently, the currents $I_{njm}(s)$ and $I_{jm}(s)$ with $I_{njm}(s) + I_{jm}(s) = I_{stim}(s)$ are transmitted across the membrane to close the circuit. The extracellular current $I_{Xsub}(s)$ responsible for generating the eFP $V_{Xsub}(s)$ at Point X equals $I_{jm}(s)$. This $I_{jm}(s)$ is minimally affected by presence of the membrane-substrate junctional seal (modeled electrically by R_{jseal}) (Guo 2019), so that the entire cell membrane can be considered to have a uniform transmembrane current density and $I_{jm}(s)$ can be approximated as $\beta_{jm}I_{stim}(s)$, where β_{jm} is the percentage of the junctional membrane area to the entire cell membrane area (see Exercise 6.4). In the absence of the electrode conductor (or the electrode is not connected to the *GND* through the amplifier's input terminal),

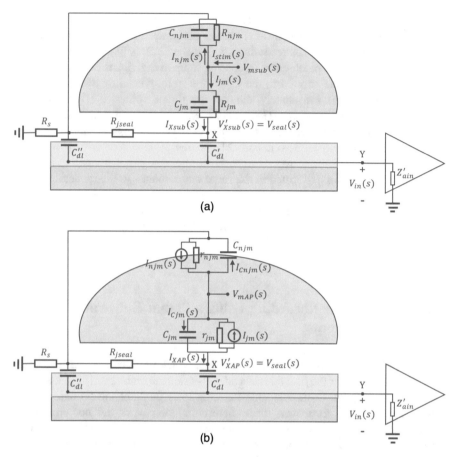

Fig. 6.3 Abstracted models of the planar microelectrode–cell membrane interfaces and their equivalent electrical circuits. (**a**) The equivalent electrical circuit during the subthreshold depolarization phase. (**b**) The equivalent electrical circuit during the AP phase. *njm* stands for nonjunctional membrane; and *jm* for junctional membrane. Adapted with permission from (Guo 2020a). Copyright © 2020 IOP Publishing Ltd

$$V_{Xsub}(s) = R_{jseal}I_{jm}(s) + R_s\left(I_{jm}(s) + I_{njm}(s)\right) = \left(\beta_{jm}R_{jseal} + R_s\right)I_{stim}(s) \quad (6.6)$$

According to $I_{stim}(s) = \left(\frac{1}{R_m} + sC_m\right)V_{msub}(s)$ from Eq. (5.3), we have

$$V_{Xsub}(s) = \left(\beta_{jm}R_{jseal} + R_s\right)\left(\frac{1}{R_m} + sC_m\right)V_{msub}(s) \quad (6.7)$$

At neuronal frequencies, Eq. (6.7) approximates to

$$V_{Xsub}(s) \approx \left(\beta_{jm}R_{jseal} + R_s\right)C_m \cdot sV_{msub}(s) \tag{6.8}$$

During the AP phase (Fig. 6.3b), the cell membrane acts as an electromotive force (*e.g.*, a battery), and the $I_{Na}(s)$ and $I_K(s)$ result from intrinsic properties of the membrane and do not depend on the presence of R_{jseal}, so that the transmembrane current density in the junctional membrane is not affected by the junctional seal. We have $I_{jm}(s) = \beta_{jm}I_{AP}(s)$, where $I_{AP}(s) = I_{jm}(s) + I_{njm}(s)$ is the overall transmembrane AP current. In this case, $I_{XAP}(s) = -I_{jm}(s)$ (Note that $I_{Cnjm}(s)$ and $I_{Cjm}(s)$ are now virtual capacitive currents as defined in Chap. 2, Sect. 2.3.3), and

$$V_{XAP}(s) = -R_{jseal}I_{jm}(s) - R_s\left(I_{jm}(s) + I_{njm}(s)\right) = -\left(\beta_{jm}R_{jseal} + R_s\right)I_{AP}(s) \tag{6.9}$$

According to $I_{AP}(s) = C_m \cdot sV_{mAP}(s)$ from Eq. (5.5), we have

$$V_{XAP}(s) = -\left(\beta_{jm}R_{jseal} + R_s\right)C_m \cdot sV_{mAP}(s) \tag{6.10}$$

From Eqs. (6.8) and (6.10), we see that during subthreshold depolarization, the eFP $V_{Xsub}(s)$ is proportional to the *first time derivative* of the transmembrane voltage $V_{msub}(s)$, while during the AP phase, the eAP $V_{XAP}(s)$ is proportional to the *negative first time derivative* of the iAP $V_{mAP}(s)$. These findings confirm the earlier basic relationships in Sect. 6.1. The presence of the junctional seal does substantially boost the amplitude of the eFP in the junction with an additional term of $\beta_{jm}R_{jseal}$ in the coefficient comparing to Eqs. (6.4) and (6.5). Fitting empirical data to the parameters in Eqs. (6.8) and (6.10) gives estimations of $v_{Xsub\_peak} = 1.9\mu V$ and $v_{XAP\_peak} = -33.7\mu V$ (see Exercise 6.6). v_{XAP_peak} is in the typical amplitude range of reported eAPs: -10 to $-500\mu V$, whereas v_{Xsub_peak} is below the noise level (v_{pp}: $10–40\mu V$) of typical commercial multielectrode array systems and thus unobservable by this class of planar microelectrodes (Xie et al. 2012; Spira and Hai 2013). Because R_{jseal} (*e.g.*, 0.1–1.2 MΩ (Dipalo et al. 2017; Hai et al. 2009)) is two to three orders of magnitude higher than R_s (~ 1 kΩ) and $\beta_{jm} \approx 0.3 – 0.5$, the amplitude of $v_{XAP}(t)$ is boosted by 30–500 times comparing to the open-field recording in Eq. (6.5) (compare Exercises 6.5 and 6.6).

When the electrode conductor is considered and is wired to the amplifier's input terminal, let us first assume that the cell covers the entire electrode surface (*i.e.*, $C''_{dl} = 0$ in Fig. 6.3), then $Z_e = \frac{1}{sC'_{dl}}$, and according to Eqs. (3.2) and (3.3),

$$V'_X(s) = \frac{Z_e + Z'_{ain}}{Z''_e + Z'_{ain}} V_X(s) \tag{6.11}$$

$$V_{in}(s) = \frac{Z'_{ain}}{Z''_e + Z'_{ain}} V_X(s) \tag{6.12}$$

where $Z''_e = Z_e + R_{jseal} + R_s$ is the *in situ electrode impedance* of the planar substrate microelectrode with the cell overlying on the microelectrode. If the cell fully covers the electrode, Z_e corresponds to its entire C_{dl} and Z''_e can be measured directly; otherwise, Z_e only corresponds to the covered C'_{dl} as depicted in Fig. 6.3, and Z''_e needs to be extracted from the direct measurement due to the shunting effect of the uncovered C''_{dl}.

When a whole-cell patch-clamp recording is performed in the same setup, *e.g.*, for comparison purpose, R_s is grounded in the vicinity of the nonjunctional membrane (*i.e.* $R_s = 0$), and Eqs. (6.11) and (6.12) become

$$V'_X(s) = \frac{1}{1 + \frac{R_{jseal}}{Z_e + Z'_{ain}}} V_X(s) \tag{6.13}$$

$$V_{in}(s) = \frac{Z'_{ain}}{Z_e + R_{jseal} + Z'_{ain}} V_X(s) \tag{6.14}$$

Interestingly, if the cell only partially covers the electrode surface in Fig. 6.3, the exposed electrode surface (modeled electrically by C''_{dl}) functions to shunt the effective input impedance of the amplifier, which attenuates the magnitude of Z'_{ain} (*i.e.*, reducing the signal-to-noise ratio (SNR) of the recording). The larger the exposed electrode area, the worse this effect. When R_s is not grounded, C''_{dl} can pick up the eFP on R_s, but this open-field eFP is far below the noise floor for detection (see Exercise 6.5). The overall effect of C''_{dl} is still a shunting effect to Z'_{ain}, as the voltage across R_s is neglectable and R_s essentially functions as a short wire to *GND*.

6.3 Optimizing the Recording Quality

When recording in an open-field electrolyte without a tight seal between the electrode and the cell membrane (Eqs. (6.4) and (6.5)), the eFP is generally unobservable, as it is below the noise level. Therefore, we use the planar microelectrode recording in Fig. 6.3 as an example to lead the following discussions, while the conclusions can be extended to many other cases where an even tighter seal is formed at the electrode–cell membrane interface, *e.g.*, the gold mushroom-shaped microelectrode (Hai et al. 2010a, b, 2009) and the nanopillar electrodes of a variety of forms (Dipalo et al. 2017; Xie et al. 2012; Robinson et al. 2012) (also see Chap. 9). Because, a planar microelectrode cannot extracellularly record subthreshold potentials (*e.g.*, EPSP and IPSP, see estimation in Sect. 6.2 and Exercise 6.6) from small mammalian neurons, we focus on the eAP recording according to Eq. (6.10). Substituting Eq. (6.10) into Eq. (6.12) and setting $s = j\omega$ for Fourier analysis, we have the overall frequency response of the LTI system between the voltage source (i.e., the input) $V_{mAP}(j\omega)$ and the recording $V_{in}(j\omega)$ (i.e., the output) as

$$H(j\omega) = \frac{V_{in}(j\omega)}{V_{mAP}(j\omega)} = -\frac{Z'_{ain}(\beta_{jm}R_{jseal} + R_s)C_m}{Z''_e + Z'_{ain}} \cdot j\omega \qquad (6.15)$$

This $H(j\omega)$ is a frequency-shaping filter that modulates the magnitude and phase of the frequency spectrum $V_{mAP}(j\omega)$ of the iAP. If the phase angle $\angle H(j\omega)$ still approximates to $-90°$, $v_{in}(t)$ maintains the negative first time derivative relationship to $v_{mAP}(t)$. Anyway, deconvolution should always be performed to recover $v_X(t)$ from $v_{in}(t)$ (Rossant et al. 2012; Rossert et al. 2009; Robinson et al. 2012) based on Eq. (6.12), which is the negative first time derivative of the iAP $v_{mAP}(t)$ according to Eq. (6.10). Next, we investigate how the parameters in Eq. (6.15) influence the SNR of the recording through affecting the magnitude $|H(j\omega)|$. Note, Z'_{ain} and Z''_e are also complex functions of $j\omega$.

6.3.1 Factors Affecting the SNR

The quality of a neural recording $r(t) = s(t) + N(t)$, where the signal $s(t)$ is deterministic and the background noise $N(t)$ is a stationary stochastic process, is commonly characterized by the SNR defined as the ratio of the power of the signal to the power of the noise:

$$SNR = \frac{P_s}{P_N} = \frac{\frac{1}{T}\int_0^T s^2(t)dt}{R_N(0)} \qquad (6.16)$$

where T is the duration of an observation interval and $R_N(\tau) = \mathcal{E}[N(t)N(t+\tau)]$ is the autocorrelation function of the noise. According to Parseval's relation (Oppenheim and Willsky 1997), $\int_0^T s^2(t)dt = \int_{-\infty}^{+\infty}|s(t)|^2 dt = \frac{1}{2\pi}\int_{-\infty}^{+\infty}|S(j\omega)|^2 d\omega$, where $s(t)$ is assumed 0 outside the interval T and $S(j\omega)$ is its Fourier transform or power spectrum density. In our particular case where $S(j\omega) = V_{in}(j\omega) = H(j\omega)V_{mAP}(j\omega)$,

$$P_s = \frac{1}{2\pi T} \int_{-\infty}^{+\infty} |H(j\omega)|^2 |V_{mAP}(j\omega)|^2 d\omega \qquad (6.17)$$

Since $V_{mAP}(j\omega)$ is the frequency spectrum of a given voltage source iAP, the magnitude $|H(j\omega)|$ directly affects P_s and thus the SNR of the recording.

In a planar MEA recording environment, sources of noises include the intrinsic thermal and shot noises of the recording circuit and electrical noises picked up by the circuits (*e.g.*, line and RF noises, which have different frequency ranges from the neural signal and thus can be filtered out from the final recording). The unremovable intrinsic thermal and shot noises are the background noise $N(t)$ left in the final recording and involved in the calculation of the SNR. Thermal noise is generated in electronics according to Ohm's law $P = I^2 R$. Thus, the electrical resistances along the electrical current pathway in the recording channel, including the resistance in the electrode material (R_e in Fig. 3.3), need to be minimized. For a planar micro-electrode, R_e is very small and thus can be neglected, as did in Fig. 6.3.

Next, let us scrutinize influences of the parameters and variables in Eq. (6.15) on the $|H(j\omega)|$:

Constant parameters. Effects of the constant parameters C_m, β_{jm}, R_s, and R_{jseal} to the eAP $V_{XAP}(j\omega)$ are shown in Eq. (6.10), and consequently, $|H(j\omega)| \propto (\beta_{jm}R_{jseal} + R_s)C_m$. R_sC_m corresponds to contribution from the overall cell membrane capacitive current through the open-field solution spreading resistance R_s. According to Eq. (6.5), R_s is insufficient to produce an observable signal above the noise level for small mammalian neurons with a small C_m. However, this is not the case for large neurons from the invertebrate such as leech (Jenkner and Fromherz 1997) and *Aplysia* (Hai et al. 2010a, b, 2009). So, larger cells with a larger C_m help to improve the SNR. However, the major contribution of C_m is taking effect through $\beta_{jm}C_mR_{jseal}$, the junctional membrane that forms a tight seal with the substrate; and as such, the larger the junctional membrane capacitance $\beta_{jm}C_m$, which determines the junctional current $I_{jm}(j\omega)$ in Fig. 6.3b, and the tighter of the membrane-substrate seal, which determines R_{jseal}, the higher the SNR (Hai et al. 2009). Accordingly, major efforts have been devoted to increase the R_{jseal} and the junctional current $I_{jm}(j\omega)$ for improved recording quality (Jenkner and Fromherz 1997; Hai et al. 2010a, b, 2009; Dipalo et al. 2017; Xie et al. 2012; Lin et al. 2014; Cohen et al. 2008; Robinson et al. 2012).

Z_e'' *and* Z_{ain}'. If the electrode recording system is properly engineered and configured, $C_{lsh} = 0$, $C_{esh} = 0$, and $R_{sc} = \infty$ in Fig. 3.3. Equation (6.12) describes how the electrode recording system affects the recording $V_{in}(j\omega)$ of the eAP $V_{XAP}(j\omega)$. Basically, Z_e'' and Z_{ain}' form a voltage-divider circuit with a frequency response of $\frac{Z_{ain}'}{Z_e''+Z_{ain}'}$. As a second-order system (Guo 2020b), this transmission process both attenuates the signal amplitude and shifts the phase. This is the reason for the common call for (1) decreasing the electrode's conventional impedance Z_e' via increasing the C_{dl} (Arcot Desai et al. 2010; Castagnola et al. 2014) and (2) increasing the amplifier's input impedance Z_{ain} (Nelson et al. 2008). However, it needs to be emphasized that it is the electrode recording impedance Z_e corresponding to the cell-covered C'_{dl} (Fig. 6.3) that needs to be decreased (thus C'_{dl} needs to be increased), whereas the uncovered C''_{dl} needs to be decreased. Merely increasing the overall C_{dl} will not necessarily improve the SNR, rather, in cases that the uncovered C''_{dl} is large, can severely diminish the SNR. Therefore, planar microelectrodes with a diameter comparable to that of the overlying cell body are preferred. Under this condition, the larger the C'_{dl}, the smaller the amplitude attenuation and phase distortion.

R_{jseal}. In Eq. (6.15), R_{jseal} appears in both the numerator and denominator. To determine how $|H(j\omega)|$ depends on R_{jseal}, we calculate $\frac{\partial |H(j\omega)|}{\partial R_{jseal}}$ and find that it is positive for all ω and that the larger $|Z'_{ain}|$, the larger its value. This means that (1) a high R_{jseal} can effectively boost the SNR and (2) a high $|Z'_{ain}|$ can further augment this effect. R_s has a similar effect, but as it is usually unchangeable in a recording environment and it is more than two orders of magnitude smaller, its effect is minimum.

6.4 Summary

- During subthreshold depolarization and hyperpolarization, the eFP is proportional to the first time derivative of the transmembrane voltage change. During the AP, the eFP or the eAP is proportional to the *negative* first time derivative of the iAP. These relationships result because the same net transmembrane current produces the $v_m(t)$ by charging the inner surface of the membrane capacitor C_m, whereas producing the $v_X(t)$ through R_s.
- Any extracellularly recorded subthreshold potentials, e.g., synaptic potentials, are also biphasic in their full cycle.
- The base width of the major iAP peak is ~1 ms, corresponding to a characteristic frequency of ~1 kHz, whereas the base width of the major eAP peak is less than 0.5 ms, corresponding to a characteristic frequency greater than 2 kHz. Also, the major eAP peak does not correspond to the major iAP peak which actually maps to the transiting zero point in the eAP.
- When recording using a planar substrate microelectrode, the presence of the junctional seal between the cell membrane and the substrate substantially boosts the amplitude of the eFP in the junction.
- When recording using a planar substrate microelectrode, if the cell only partially covers the electrode surface, the exposed electrode surface functions to shunt the effective input impedance of the amplifier, deteriorating the SNR of the recording. The larger the exposed electrode area, the worse this deteriorative effect.
- The recorded voltage $V_{in}(j\omega)$ is related to the eFP $V_X(j\omega)$ by Eq. (6.12). This is the reason for the common call for (1) decreasing the electrode's conventional impedance Z'_e via increasing the C_{dl} and (2) increasing the amplifier's input impedance Z_{ain}. However, it needs to be emphasized that it is the electrode recording impedance Z_e corresponding to the cell-covered C'_{dl} (Fig. 6.3) that needs to be decreased (thus C'_{dl} needs to be increased), whereas the uncovered C''_{dl}

needs to be decreased. Merely increasing the overall C_{dl} will not necessarily improve the SNR, rather, in cases that the uncovered C''_{dl} is large, can severely diminish the SNR. Therefore, planar microelectrodes with a diameter comparable to that of the overlying cell body are preferred. The selection of Z_{ain} is advised to be at least one order of magnitude higher than Z'_e.

Note: This chapter was adapted from (Guo 2020a) with permission.

Exercises

6.1 For extracellular neural recording,

(a) What is the effect of the electrode recording system to the eFP?
(b) How is the eFP $V_X(s)$ produced by the net outward transmembrane current?
(c) What is the relationship between the $V_X(s)$ and the $V_m(s)$?
(d) What is the relationship between the recording $V_{in}(s)$ and the $V_X(s)$?

6.2 Referring to Fig. 6.2,

(a) How is the eFP related to the AC transmembrane voltage?
(b) If an EPSP can be detected extracellular, how does the recording look like?
(c) Why is the eAP's major peak negative? Which part of the iAP does it correspond to? Which part of the eAP does the iAP peak correspond to?
(d) What is the characteristic frequency of the eAP? For an extracellular recording microelectrode, do you think it is appropriate to only measure its impedance at 1 kHz?

6.3 Referring to Fig. 6.2b, for extracellular neural recording,

(a) Why is the first positive peak in the eFP sometimes not detected?
(b) Why is the amplitude of the eAP so low, comparing to that of the iAP?
(c) How to boost the amplitude of the eAP?

6.4 Without considering the electrode conductor, prove that the membrane-substrate junctional seal exerts a minimum effect on $I_{jm}(s)$ in Fig. 6.3a, so that the entire cell membrane can be considered to have a uniform transmembrane current density.

6.5 A spherical neuron of 7.5μm radius is suspended in an infinite, homogenous PBS solution. The specific membrane capacitance is 0.01 pF/μm². The resistivity of the PBS solution is 0.09 Ω·m at room temperature. The maximum rising slope of a subthreshold depolarization is $\left[\frac{\Delta v_{msub}}{\Delta t}\right]_{max} = \frac{10\text{ mV}}{1\text{ ms}} = 10$ V/s; and the maximum rising slope of an AP is $\left[\frac{\Delta v_{mAP}}{\Delta t}\right]_{max} = \frac{90\text{ mV}}{0.3\text{ ms}} = 300$ V/s.

(a) Find the spreading resistance R_s from the outer cell surface to the infinity where the *GND* is placed.
(b) What is the amplitude of the eFP $v_{Xsub}(t)$ on the outer cell surface during the subthreshold depolarization?

    (c) What is the amplitude of the eAP $v_{XAP}(t)$ on the outer cell surface during the AP?

    (d) Can the $v_{Xsub}(t)$ and $v_{XAP}(t)$ be detected by a conventional microelectrode system?

6.6 As shown in Fig. 6.3, a neuron attaches onto the surface of a planar microelectrode. The junctional membrane capacitance $C_{jm} = 1.77$ pF, the nonjunctional membrane capacitance $C_{njm} = 3.53$ pF, and the total cell membrane capacitance $C_m = 5.3$ pF. The junctional seal resistance $R_{seal} = 0.1$ MΩ. R_s, $\left[\frac{\Delta v_{msub}}{\Delta t}\right]_{max}$, and $\left[\frac{\Delta v_{mAP}}{\Delta t}\right]_{max}$ are the same as in Exercise 6.5.

    (a) What is the amplitude of the eFP $v_{Xsub}(t)$ at Point X during the subthreshold depolarization?

    (b) What is the amplitude of the eAP $v_{XAP}(t)$ at Point X during the AP?

    (c) Can the $v_{Xsub}(t)$ and $v_{XAP}(t)$ be detected by the planar microelectrode system shown?

    (d) Comparing to the solutions to Exercise 6.5, what conclusion can you reach?

6.7 What are the key factors effecting the eAP recording? How should they be controlled in order to obtain a recording with a high SNR?

References

Arcot Desai S, Rolston JD, Guo L, Potter SM (2010) Improving impedance of implantable microwire multi-electrode arrays by ultrasonic electroplating of durable platinum black. Frontiers in neuroengineering 3:5

Castagnola E, Ansaldo A, Maggiolini E, Ius T, Skrap M, Ricci D, Fadiga L (2014) Smaller, softer, lower-impedance electrodes for human neuroprosthesis: a pragmatic approach. Frontiers in neuroengineering 7:8

Cohen A, Shappir J, Yitzchaik S, Spira ME (2008) Reversible transition of extracellular field potential recordings to intracellular recordings of action potentials generated by neurons grown on transistors. Biosensors & Bioelectronics 23 (6):811-819. doi:https://doi.org/10.1016/j.bios.2007.08.027

Dipalo M, Amin H, Lovato L, Moia F, Caprettini V, Messina GC, Tantussi F, Berdondini L, De Angelis F (2017) Intracellular and Extracellular Recording of Spontaneous Action Potentials in Mammalian Neurons and Cardiac Cells with 3D Plasmonic Nanoelectrodes. Nano Lett 17 (6):3932-3939. doi:https://doi.org/10.1021/acs.nanolett.7b01523

Fromherz P (2002) Electrical interfacing of nerve cells and semiconductor chips. Chemphyschem 3 (3):276-284. doi: https://doi.org/10.1002/1439-7641(20020315)3:3<276::Aid-Cphc276>3.0.Co;2-A

Grattarola M, Martinoia S (1993) Modeling the neuron-microtransducer junction: from extracellular to patch recording. IEEE transactions on bio-medical engineering 40 (1):35-41. doi:https://doi.org/10.1109/10.204769

Guo L (2019) On neural recording using nanoprotrusion electrodes. J Neural Eng

Guo L (2020a) Perspectives on electrical neural recording: a revisit to the fundamental concepts. J Neural Eng 17 (1):013001. doi:https://doi.org/10.1088/1741-2552/ab702f

Guo L (2020b) Principles of functional neural mapping using an intracortical ultra-density micro-electrode array (ultra-density MEA). J Neural Eng 17. doi:https://doi.org/10.1088/1741-2552/ab8fc5

Hai A, Dormann A, Shappir J, Yitzchaik S, Bartic C, Borghs G, Langedijk JPM, Spira ME (2009) Spine-shaped gold protrusions improve the adherence and electrical coupling of neurons with the surface of micro-electronic devices. Journal of the Royal Society Interface 6 (41):1153-1165. doi:https://doi.org/10.1098/rsif.2009.0087

Hai A, Shappir J, Spira ME (2010a) In-cell recordings by extracellular microelectrodes. Nature methods 7 (3):200-U250. doi:https://doi.org/10.1038/Nmeth.1420

Hai A, Shappir J, Spira ME (2010b) Long-Term, Multisite, Parallel, In-Cell Recording and Stimulation by an Array of Extracellular Microelectrodes. Journal of neurophysiology 104 (1):559-568. doi:https://doi.org/10.1152/jn.00265.2010

Hierlemann A, Frey U, Hafizovic S, Heer F (2011) Growing Cells Atop Microelectronic Chips: Interfacing Electrogenic Cells In Vitro With CMOS-Based Microelectrode Arrays. P Ieee 99 (2):252-284. doi:https://doi.org/10.1109/Jproc.2010.2066532

Horch KW, Dhillon GS (2004) Neuroprosthetics: theory and practice. World Scientific,

Jenkner M, Fromherz P (1997) Bistability of membrane conductance in cell adhesion observed in a neuron transistor. Phys Rev Lett 79 (23):4705-4708. doi:https://doi.org/10.1103/PhysRevLett.79.4705

Lin ZLC, Xie C, Osakada Y, Cui Y, Cui BX (2014) Iridium oxide nanotube electrodes for sensitive and prolonged intracellular measurement of action potentials. Nat Commun 5. doi: https://doi.org/10.1038/ncomms4206

Nelson MJ, Pouget P, Nilsen EA, Patten CD, Schall JD (2008) Review of signal distortion through metal microelectrode recording circuits and filters. J Neurosci Meth 169 (1):141-157. doi:https://doi.org/10.1016/j.jneumeth.2007.12.010

Oppenheim AV, Willsky AS (1997) Signals and Systems. 2nd Edition edn. Prentice Hall,

Robinson JT, Jorgolli M, Shalek AK, Yoon MH, Gertner RS, Park H (2012) Vertical nanowire electrode arrays as a scalable platform for intracellular interfacing to neuronal circuits. Nature nanotechnology 7 (3):180-184. doi:https://doi.org/10.1038/nnano.2011.249

Rossant C, Fontaine B, Magnusson AK, Brette R (2012) A calibration-free electrode compensation method. Journal of neurophysiology 108 (9):2629-2639. doi:https://doi.org/10.1152/jn.01122.2011

Rossert C, Straka H, Glasauer S, Moore LE (2009) Frequency-domain analysis of intrinsic neuronal properties using high-resistant electrodes. Frontiers in Neuroscience 3. doi: https://doi.org/10.3389/neuro.17.002.2009

Spira ME, Hai A (2013) Multi-electrode array technologies for neuroscience and cardiology. Nature nanotechnology 8 (2):83-94

Xie C, Lin ZL, Hanson L, Cui Y, Cui BX (2012) Intracellular recording of action potentials by nanopillar electroporation. Nature nanotechnology 7 (3):185-190. doi:https://doi.org/10.1038/Nnano.2012.8

Chapter 7
Extracellular Recording of Propagating Action Potentials

In most extracellular recordings, we record the signals generated by the soma, because conventional microelectrodes cannot detect the signals produced by the fine axons and dendrites of mammalian neurons, unless they attach onto the electrode's surface (Guo 2020a, b). In our previous analyses, we treated the neuronal soma as a homogeneous monopole current source and ignored any propagation of the transmembrane voltage change across the soma. This treatment is feasible when the microelectrode size is comparable or larger than the soma size, as is the case for most extracellular recording situations. However, when the electrode size is substantially smaller than that of the neuronal soma, the propagation effect has to be taken into account for a better accuracy. One particular and common example is the in vitro recording of cardiomyocytes, which can spread to an area of hundreds of micrometers in diameter, using substrate-integrated micro/nanoelectrodes. In substrate-cultured cardiomyocytes, the AP spreads as a two-dimensional plane wave, but to a microelectrode underneath, the electrode only sees a one-dimensional plane wave. Another case in which the propagation effect needs to be considered is the extracellular recording along a neuronal axon using nanoelectrode arrays (Patolsky et al. 2006; Bakkum et al. 2013). Again, the electrode sees a one-dimensional wave. Thus, in this chapter, we analyze how the propagation of a one-dimensional wave affects the eFP recorded by an underneath micro/nanoelectrode.

7.1 AP Propagation and Its Modeling

Due to the refractory property of neuronal membrane, when an AP is initiated at one point, it can only propagate away from that point, but cannot come backward. Referring to Fig. 7.1, during the AP at Point 2, the AP current $i_{2AP}(t)$ spreads bilaterally ($i_2'(t)$ and $i_2''(t)$) along the inner membrane around Point 2, so that the actual AP slopes and peaks at Point 2 are slightly attenuated. It is such bilaterally

L. Guo, *Principles of Electrical Neural Interfacing*,
https://doi.org/10.1007/978-3-030-77677-0_7

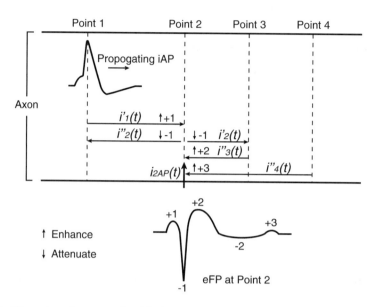

Fig. 7.1 Illustration of a propagating AP along an axon for analyzing the eFP

spreading currents that produce (1) an intracellular ionic current wave front $i'_2(t)$ at the succeeding membrane at Point 3 to depolarize the membrane for the subsequent initiation of AP there and (2) an intracellular ionic current wave tail $i''_2(t)$ at the proceeding membrane at Point 1 where an AP cannot be reinitiated due to the membrane's refractory property. The distance such a lateral ionic current can spread along a passive unmyelinated axon is conventionally characterized as the spatial constant τ, and the transmembrane voltage along this distance decays exponentially as a result of the first-order RC circuit formed by the parallel C_m and R_m (Johnston and Wu 1995). When $i'_2(t)$ and $i''_2(t)$ exit the axonal membrane capacitively at Points 3 and 1, respectively, they flow along the extracellular membrane back to the extracellular space around Point 2 to close the circuit and attenuate the first negative peak marked as "−1" on the eFP at Point 2.

7.1.1 Forward Propagating Intracellular Current

When Point 1 fires an AP, its $i'_1(t)$ spreads to Point 2 for depolarization. This $i'_1(t)$ is our previously discussed $i_{stim}(t)$ during the subthreshold analyses in Chap. 6. As shown in Fig. 7.1, it generates the first positive peak marked as "+1" on the eFP at Point 2 according to Eq. (6.1) (Fig. 6.2a). Because the continuing depolarization at Point 2 by $i'_1(t)$ is suddenly interrupted by the onset of an AP at Point 2 when the threshold is exceeded, the latter stages of this subthreshold transmembrane voltage change is simply overwhelmed in the AP, but nonetheless, the full dynamics of this subthreshold voltage change still exists and superimposes to the AP at Point 2.

7.1.2   Backward Propagating Intracellular Current

When the AP propagates to Point 3, its $i_3''(t)$ flows back to Point 2 to depolarize the membrane there. Suppose the distance between Points 2 and 3 is close enough so that the AP at Point 2 is in its repolarization phase. Thus, $i_3''(t)$ counteracts the repolarization at Point 2 and physically slows down the repolarization at Point 2. According to Eq. (6.2), the second positive peak marked as "+2" on the eFP at Point 2 should be attenuated. However, this $i_3''(t)$ does not attenuate the outward $i_K(t)$ flowing into the extracellular space at Point 2, rather, it exits the membrane through the already opened K^+ channels and strengthens the outward $i_K(t)$ to create a more prominent overshoot in the second positive peak of the eFP.

Point 4 is farther away from Point 2. When the AP fires at Point 4, the AP at Point 2 is finishing its refractory period, $i_4''(t)$ simply depolarizes the now passive membrane at Point 2 according to Eq. (6.1), producing the third positive peak marked as "+3" on the eFP at Point 2.

Therefore, the overall backward propagation effect on the eFP at Point 2 by the forward-moving AP passing Point 2 is a prolonged and enhanced second positive peak followed by a minor third positive peak in the eFP, as shown in Fig. 7.1.

7.1.3   Overall Propagating Effect on eFP

Based on the above analyses, the overall current involved in the charging of the inner membrane at Point 2 is the spatiotemporal superposition of $i_1'(t) + \left[i_{2AP}(t) - i_2'(t) - i_2''(t)\right] + i_3''(t) + i_4''(t)$. $i_1'(t)$ and $i_4''(t)$ work to produce the eFP according to Eq. 6.1, while $\left[i_{2AP}(t) - i_2'(t) - i_2''(t)\right]$ and $i_3''(t)$ work according to Eq. (6.2). Therefore, the resulted overall eFP has an asymmetric M-shaped waveform, with the first phase ("+1") in Fig. 7.1 corresponding to the first time derivative of the transmembrane voltage $v_{m2}(t)$ (Eq. 6.4), the second phase (including "−1", "+2", and "−2") corresponding to the negative first time derivative of $v_{m2}(t)$ (Eq. (6.5)), and the third phase ("+3") again corresponding to the first time derivative of $v_{m2}(t)$ (Eq. (6.4)). When the wave function is employed for a qualitative analysis, it is suggested that the eFP is approximately proportional to the second spatial derivative of $v_m(t, x)$ (Johnston and Wu 1995).

7.2   The Recorded eFP

While the full waveform of an eFP $v_X(t)$ at a point along a propagating membrane has three minor positive peaks and one major and one minor negative peaks (Fig. 7.1), according to Eq. 3.3, the recorded signal $v_{in}(t)$ may not show up the first ("+1") and third ("+3") positive and the second negative ("−2") peaks, as they are below the noise floor in $v_{in}(t)$ for conventional microelectrode recording systems. Thus, the commonly recorded eFPs of propagating APs show up with one major negative peak ("−1") and one minor positive peak ("+2"). However, using advanced nanoprotrusion electrodes together with a high input-impedance amplifier (see Chap. 9), the full waveform of such multiphasic eFPs can be revealed.

7.3   Summary

- The overall backward propagation effect on the eFP by the forward-moving AP is a prolonged and enhanced second positive peak followed by a minor third positive peak in the eFP.
- The resulted overall eFP has an asymmetric M-shaped waveform, with the first phase corresponding to the first time derivative of the transmembrane voltage, the second phase corresponding to the negative first time derivative, and the third phase again corresponding to the first time derivative.

Exercise
7.1 How is the eFP profile affected by the propagation of an AP along a membrane? Can these effects be detected in the recording of the eFP?

References

Bakkum DJ, Frey U, Radivojevic M, Russell TL, Müller J, Fiscella M, Takahashi H, Hierlemann A (2013) Tracking axonal action potential propagation on a high-density microelectrode array across hundreds of sites. Nat Commun 4 (1):1-12

Guo L (2020a) Perspectives on electrical neural recording: a revisit to the fundamental concepts. J Neural Eng 17 (1):013001. doi:https://doi.org/10.1088/1741-2552/ab702f

Guo L (2020b) Principles of functional neural mapping using an intracortical ultra-density microelectrode array (ultra-density MEA). J Neural Eng 17. doi:https://doi.org/10.1088/1741-2552/ab8fc5

Johnston D, Wu SM-S (1995) Foundations of cellular neurophysiology. MIT Press, Cambridge, MA.

Patolsky F, Timko BP, Yu G, Fang Y, Greytak AB, Zheng G, Lieber CM (2006) Detection, stimulation, and inhibition of neuronal signals with high-density nanowire transistor arrays. Science 313 (5790):1100-1104

Chapter 8
Recording Using Field-Effect Transistors

While the majority of electrical neural recordings are performed using electrodes, there is another class of electrical biopotential sensors using field-effect transistors (FETs), which have been used for both intracellular (Zhao et al. 2019; Tian et al. 2010) and extracellular (Patolsky et al. 2006; Fromherz 2002; Hierlemann et al. 2011; Park et al. 2019) neuronal recordings. Such FET biopotential sensors use the gate as the electrode to sense the potential. The advantages of such an approach include the integrated FET amplifier for direct signal amplification to improve the recording SNR and high planar sensor packing density leveraging the complementary metal oxide semiconductor (CMOS) technology.

Figure 8.1 depicts the equivalent electrical circuit for a metal oxide semiconductor FET (MOSFET) neuronal potential sensor in an extracellular configuration, where the neuron sits on top of the gate. The MOSFET itself is modeled using the small-signal "π" model (Sedra and Smith 2014) (Fig. 8.1b), which is a voltage-controlled current source. The gate itself is modeled as a capacitor C_{gs} between G and S. And the gate–electrolyte interface is modeled as a conventional EDL capacitor C_{dl}, in series to the gate capacitor C_{gs}. This MOSFET recording circuit is essentially the same voltage-divider circuit as in Fig. 3.3b with the $Z_{ain} = \frac{1}{sC_{gs}}$, and the extracellular recording configuration is the same as in Fig. 6.3. Thus, we have the eFP in the presence of the FET as

$$V'_X(s) = \frac{Z_e + Z_{ain}}{Z''_e + Z_{ain}} V_X(s) \tag{8.1}$$

And the recording of the FET as

$$V_{in}(s) = V_{gs}(s) = \frac{Z_{ain}}{Z''_e + Z_{ain}} V_X(s) \tag{8.2}$$

where $Z_e = \frac{1}{sC_{dl}}$ and the in situ electrode impedance $Z''_e = Z_e + R_{jseal} + R_s$.

L. Guo, *Principles of Electrical Neural Interfacing*,
https://doi.org/10.1007/978-3-030-77677-0_8

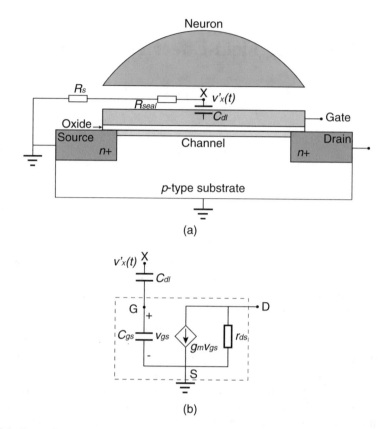

Fig. 8.1 Extracellular neuronal recording using a MOSFET. (**a**) The neuron is placed on top of the gate using a micropatterning technique. (**b**) The small-signal "π" model (dashed box) of the MOSFET is used for the AC analysis

To have a good recording SNR, we need Z_e to be at least one order of magnitude smaller than Z_{ain}, which requires the C_{dl} to be at least one order of magnitude larger than the C_{gs}. According to $C = \varepsilon_0 \varepsilon_r \frac{A}{d}$, as the geometric area A and the dielectric distance d of C_{dl} and C_{gs} are almost the same (both are a few nanometers) and the relative permittivity of PBS is 79.0 and that of SiO_2 is 3.9 at room temperature, C_{dl} is more than 20 times larger than C_{gs}. The requirement is met nicely, which explains why the recording SNR using a FET is generally satisfactory comparing to using a conventional planar microelectrode. Furthermore, the gate conductor can be made larger than the underneath channel area, and the smooth metal surface can be coated with a rough material such as platinum black or conducting polymer, to further increase the C_{dl}; but this is not necessary, as the benefit gained is diminishing.

Another contributor to the better recording quality using a FET is that the recording is directly amplified by the FET itself, thus avoiding noise contamination and signal attenuation along a transmission lead.

Therefore, our analysis well explains the impressive SNR performance of neuronal recording using a FET over a conventional microelectrode.

Summary

- Neuronal recording using a FET also forms a voltage-divider circuit, where the amplifier's input impedance is $Z_{ain} = \frac{1}{sC_{gs}}$.
- The EDL impedance of the gate is more than 20 times smaller than the input impedance of the FET amplifier; and according to Eq. (8.2), FET neuronal recording thus features a high SNR.
- A further contributor to the better recording quality using a FET is that the recording is directly amplified by the FET itself, thus avoiding noise contamination and signal attenuation along a transmission lead.

Exercises

8.1 What are the advantages and disadvantages of neural recording using a FET-based sensor?

8.2 Why does FET-based neural recording feature a high SNR?

8.3 Can FET-based circuits be used for neural stimulation? Can electroporation to the cell membrane be used to enhance the extracellular recording?

References

Fromherz P (2002) Electrical interfacing of nerve cells and semiconductor chips. Chemphyschem 3 (3):276-284. doi:https://doi.org/10.1002/1439-7641(20020315)3

Hierlemann A, Frey U, Hafizovic S, Heer F (2011) Growing Cells Atop Microelectronic Chips: Interfacing Electrogenic Cells In Vitro With CMOS-Based Microelectrode Arrays. P IEEE 99 (2):252-284. doi:https://doi.org/10.1109/Jproc.2010.2066532

Park JS, Grijalva SI, Jung D, Li S, Junek GV, Chi T, Cho HC, Wang H (2019) Intracellular cardiomyocytes potential recording by planar electrode array and fibroblasts co-culturing on multi-modal CMOS chip. Biosensors and Bioelectronics 144:111626

Patolsky F, Timko BP, Yu G, Fang Y, Greytak AB, Zheng G, Lieber CM (2006) Detection, stimulation, and inhibition of neuronal signals with high-density nanowire transistor arrays. Science 313 (5790):1100-1104

Sedra AS, Smith KC (2014) Microelectronic Circuits. The Oxford Series in Electrical and Computer Engineering, 7th edn. Oxford University Press,

Tian BZ, Cohen-Karni T, Qing Q, Duan XJ, Xie P, Lieber CM (2010) Three-Dimensional, Flexible Nanoscale Field-Effect Transistors as Localized Bioprobes. Science 329 (5993):830-834. doi: https://doi.org/10.1126/science.1192033

Zhao Y, You SS, Zhang A, Lee J-H, Huang J, Lieber CM (2019) Scalable ultrasmall three-dimensional nanowire transistor probes for intracellular recording. Nature nanotechnology 14 (8):783-790

Chapter 9
Neural Recording Using Nanoprotrusion Electrodes

Neural recording has been an intense subject of study for a long time. This yet immature field is experiencing an unprecedented resurgence in recent years thanks to worldwide advocations on brain research and neurotechnologies. Among several exciting new research directions, scaling the neural electrodes down to nanoscale to minimize trauma and immune responses has proven a major strategy in addressing the chronic reliability issue of neural implants (Wang et al. 2018). Additionally, nano neural electrodes, often adopting a nanoprotrusion structure, also offer exciting new opportunities including recording subthreshold potentials extracellularly and smoothly switching between extracellular and intracellular recording modes (Dipalo et al. 2017; Xie et al. 2012; Lin et al. 2014; Tian et al. 2010; Robinson et al. 2012). This chapter focuses on this special class of neural recording electrodes that has only emerged in the past decade or so. This class of nanoelectrodes has a variety of names and generally features as protruded nanopillars on a substrate, on top of which the neurons or cardiomyocytes are cultured (Fig. 9.1). Such a configuration produces a very tight seal between the nanoprotrusion and the wrapping cell membrane, enabling the extracellular recording of subthreshold transmembrane voltage changes which cannot be detected by conventional microelectrodes (see Chap. 6, Sect. 6.2). In addition, this type of nanoprotrusion electrodes can also record "intracellular-like" APs after porating the membrane using a physical stimulus such as a high voltage or a laser pulse.

While intracellular recordings can obtain a unique waveform of the AP, extracellular recordings are complicated by the location, size, and shape of the electrode, as well as neighboring neural structures (Grattarola and Martinoia 1993; Gold et al. 2006; Lewandowska et al. 2015). Fortunately, the recording environment of nanoprotrusion electrodes is a very particular experimental situation that is well isolated from interferences in the surrounding macro cell-electrolyte environment, and the size and shape of these electrodes, which often adopt a nanoprotrusion structure, are relatively consistent. Furthermore, as it takes less than 1/1000 of its duration for the AP to pass over the nanojunctional membrane area, we can assume the macro membrane surrounding the nanoprotrusion electrode to be spatially

L. Guo, *Principles of Electrical Neural Interfacing*, https://doi.org/10.1007/978-3-030-77677-0_9

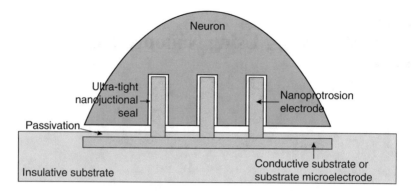

Fig. 9.1 Illustration of a typical nanoprotrusion electrode–cell interface. An array of multiple nanoprotrusion electrodes fabricated on a passivated planar microelectrode is shown. A neuron sits on top of the array, and each nanoprotrusion forms an ultra-tight nanojunctional seal with the cellular membrane, where the nanoelectrode is fully engulfed by the cell

uniform and excited in synchrony and thus can ignore the propagation effect of the AP on the recording. This makes it possible to derive a unique solution to this particular recording situation. Based on the theoretical framework in Chaps. 3 and 6, in the following, we analyze the extracellular and intracellular recordings using this class of nanoprotrusion electrodes.

9.1 Extracellular Recording by Nanoprotrusion Electrodes

First, we consider the extracellular recording using a single nanoprotrusion electrode, as illustrated in Fig. 9.2. We first consider that the substrate, which holds the nanoprotrusion electrode, is passivated (Xie et al. 2012; Robinson et al. 2012). With presence of the connected electrode recording circuit, Eqs. (3.2) and (3.3) are substantiated as

$$V'_X(s) = \frac{1}{1 + \frac{R'_s}{Z_e + Z_{in}}} V_X(s) \tag{9.1}$$

$$V_{in}(s) = \frac{Z_{in}}{Z''_e + Z_{in}} V_X(s) \tag{9.2}$$

where $R'_s = R_{njseal} + R_{jseal}$, and $Z''_e = Z_e + R'_s$ is the in situ electrode impedance of the nanoprotrusion electrode measured with the electrolyte at the exterior of the nonjunctional membrane grounded, to differentiate it from the conventional Z'_e measured without the tight membrane seals.

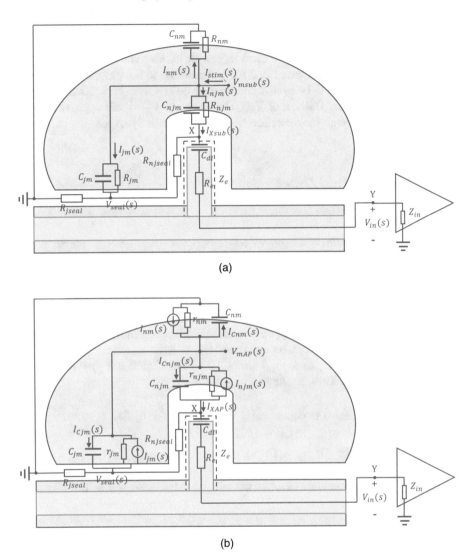

Fig. 9.2 Abstracted models of the nanoprotrusion electrode–cell membrane interfaces and their equivalent electrical circuits for extracellular recording. (**a**) During subthreshold depolarization. (**b**) During AP. Subscripts: *nm*, nonjunctional membrane; *jm*, junctional membrane; *njm*, nanojunctional membrane; *jseal*, junctional seal; *njseal*, nanojunctional seal; *stim*, stimulation; *X*, extracellular recording point; *sub*, subthreshold depolarization phase; *AP*, AP phase; *dl*, electric double layer. In (**b**), the opened ion channel resistances r_{nm}, r_{jm} and r_{njm} are treated as the internal resistances of the respective transmembrane current sources, and $I_{Cnm}(s)$, $I_{Cjm}(s)$, and $I_{Cnjm}(s)$ are virtual capacitive transmembrane currents. Adapted with permission from (Guo 2019). Copyright © 2019 IOP Publishing Ltd

Unique to the nanoprotrusion electrode's extracellular recording situation, the electrode's recording impedance Z_e is high (e.g., 54 MΩ at 1 kHz (Xie et al. 2012)) due to the nano dimensions of the electrode, and the in situ serial resistance R'_s is very high (e.g., >900 MΩ (Robinson et al. 2012)). According to Eq. (9.2), the amplifier's input impedance Z_{in} has to be even higher (e.g., 1 GΩ (Nelson et al. 2008)) in order to pick up a substantial fraction of $V_X(s)$. In this case, $V'_X(s)$ still approximates to $V_X(s)$ (Eq. (9.1)), though the distortion becomes substantially larger.

9.1.1 Subthreshold Depolarization Phase

During the subthreshold depolarization phase (Fig. 9.2a), the overall equivalent outward transmembrane current $I_{Esub}(s) = I_{nm}(s) + I_{jm}(s) + I_{njm}(s) = I_{stim}(s)$ is now split into three portions corresponding to those of the cell membrane. Assuming that the nanoprotrusion does not distort passive electrical properties of the membrane and that the membrane is spatially uniform, we define the membrane area ratios $A_{nm} : A_{jm} : A_{njm} : A_m = \beta_{nm} : \beta_{jm} : \beta_{njm} : 1$. The transmembrane current $I_{njm}(s)$ is minimally affected by the nanojunctional seal (modeled electrically by R_{njseal}), so that the entire cell membrane can be considered to have a uniform transmembrane current density and $I_{njm}(s)$ can be approximated as $\beta_{njm}I_{stim}(s)$ (see Exercise 9.1). And the portion of $I_{Esub}(s)$ exiting the nanojunctional membrane is $I_{Xsub}(s) = \beta_{njm}I_{stim}(s)$. Thus, we have

$$
\begin{aligned}
V_{Xsub}(s) &= R_{njseal}I_{njm}(s) + R_{jseal}\left(I_{jm}(s) + I_{njm}(s)\right) \\
&= \left[\beta_{njm}R_{njseal} + \left(\beta_{jm} + \beta_{njm}\right)R_{jseal}\right]I_{stim}(s) \\
&\approx \left[\beta_{njm}R_{njseal} + \left(\beta_{jm} + \beta_{njm}\right)R_{jseal}\right]C_m \cdot sV_{msub}(s)
\end{aligned}
\tag{9.3}
$$

Therefore, as in Eqs. (6.4) and (6.8), $V_{Xsub}(s)$ is proportional to the first time derivative of $V_{msub}(s)$.

9.1.2 AP Phase

The equivalent electrical circuit during the AP phase is shown in Fig. 9.2b. Assuming that the nanoprotrusion does not distort the distributions of transmembrane ion channels, as the cell membrane functions similarly to a self-perpetuated battery, the transmembrane current density in the nanojunctional membrane is not affected by the nanojunctional seal R_{njseal}, and we have $I_{njm}(s) = \beta_{njm}I_{AP}(s)$, where $I_{AP}(s) = I_{nm}(s) + I_{jm}(s) + I_{njm}(s)$ is the overall transmembrane AP current. The equivalent outward transmembrane current passing Point X is $I_{XAP}(s) = -I_{njm}(s)$. Note that $I_{Cnjm}(s)$ is a virtual capacitive current. Flowing a similar analysis, we have

$$V_{XAP}(s) = -\left[\beta_{njm}R_{njseal} + \left(\beta_{jm} + \beta_{njm}\right)R_{jseal}\right]C_m \cdot sV_{mAP}(s) \qquad (9.4)$$

Thus, $V_{XAP}(s)$ is proportional to the negative first time derivative of the iAP $V_{mAP}(s)$, as in Eqs. (6.5) and (6.10).

However, it is possible for the nanoprotrusion to distort the distributions of transmembrane ion channels, as well as changing the local membrane capacitance. These nonlinear distortions could cause the current densities across the nanojunctional membrane to deviate from those across an intact membrane, resulting in distortion to the waveforms and reduction to the amplitudes of $i_{njm}(t)$. Consequently, $v_X(t)$ could be slightly distorted and attenuated.

9.2 Recording by Nanoprotrusion Electrodes After Membrane Poration

Next, let us consider the recording of an "intracellular-like" AP by the nanoprotrusion electrode after electro- or optoporation of the nanojunctional membrane (Dipalo et al. 2017; Xie et al. 2012; Lin et al. 2014; Robinson et al. 2012). Electrically, the cytosol is now connected to the electrolyte in the nanojunction through the nanopores, which are modeled as a lumped resistor R_p as illustrated in Fig. 9.3. To the whole cell, the value of R_p (~2 GΩ) is significant due to its nanoscale cross-sectional area (Robinson et al. 2012). These nanopores thus provide a resistive leaking path to the intracellular currents, which slightly diminishes the iAP $V_m(s)$ in the intact cell to $V'_m(s)$.

9.2.1 Subthreshold Depolarization Phase

During the subthreshold depolarization phase (Fig. 9.3a), the equivalent outward transmembrane current $I_{Esub}(s) = I_{stim}(s)$ is still split into three parts corresponding to three parts of the cell membrane: $I_{nm}(s)$ and $I_{jm}(s)$ as before, and a resistive current $I'_{njm}(s)$ through R_p. Solving this circuit (Guo 2019) gives an analytical solution:

$$V_{Xsub}(s) = V'_{msub}(s) - \dfrac{\dfrac{R_p}{R_{jseal}+R_p+R_{njseal}}}{s\dfrac{C_{jm}R_{jseal}\left(R_p+R_{njseal}\right)}{R_{jseal}+R_p+R_{njseal}} + 1} V'_{msub}(s) \qquad (9.5)$$

The second term corresponds to the voltage drop across the R_p. The modulating factor is a first-order lowpass filter with a passband gain $G = \frac{R_p}{R_{jseal}+R_p+R_{njseal}}$ (e.g., 0.7)

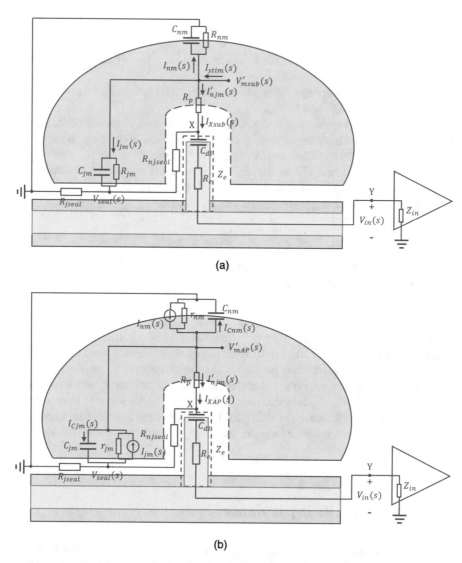

Fig. 9.3 Abstracted models of the nanoprotrusion electrode–cell membrane interfaces and their equivalent electrical circuits for intracellular-like recording after membrane poration. (a) During subthreshold depolarization. (b) During AP. Subscript: p, porated. X is the extracellular point where the potential is investigated. In (b), the opened ion channel resistances r_{nm} and r_{jm} are treated as the internal resistances of the respective transmembrane current sources, and $I_{Cnm}(s)$ and $I_{Cjm}(s)$ are virtual capacitive transmembrane currents. Adapted with permission from (Guo 2019). Copyright © 2019 IOP Publishing Ltd

and a -3-dB cutoff frequency $f_c = \dfrac{R_{jseal} + R_p + R_{njseal}}{2\pi C_{jm} R_{jseal} \left(R_p + R_{njseal} \right)} \approx \dfrac{1}{2\pi C_{jm} R_{jseal}}$ (e.g., 76 kHz) (Oppenheim and Willsky 1996). Because the frequency spectrum (<1 kHz) of $V'_{msub}(s)$ falls within its passband, this filter simply reduces to a scaling factor of G. Thus, Eq. (9.5) becomes

$$V_{Xsub}(s) = \frac{R_{jseal} + R_{njseal}}{R_{jseal} + R_p + R_{njseal}} V'_{msub}(s) \qquad (9.6)$$

Equation (9.6) is simply a voltage-divider circuit, indicating that no current is flowing across the junctional membrane (i.e., $I_{jm}(s) = 0$).

9.2.2   AP Phase

During the AP phase (Fig. 9.3b), the nonjunctional and junctional membranes generate the AP current $I'_{AP}(s) = I_{nm}(s) + I_{jm}(s)$, which generates the iAP $V'_{mAP}(s)$ by charging the $C_{nm} + C_{jm}$ after deducting the leakage $I'_{njm}(s)$ through R_p. It is noted that $V'_{mAP}(s)$ is slightly lower than $V_{mAP}(s)$ when the nanojunctional membrane is intact, as the total current charging $C_{nm} + C_{jm}$ is smaller than $I'_{AP}(s)$ due to the leakage, though the leaking current $I'_{njm}(s)$ is minute and minimally affects the $V_{mAP}(s)$. Solving this circuit gives

$$V_{XAP}(s) \approx \frac{R_{njseal}}{R_p + R_{njseal}} V'_{mAP}(s) - \frac{R_p}{R_p + R_{njseal}} R_{jseal} C_{jm} \cdot s V'_{mAP}(s) \qquad (9.7)$$

Thus, the eAP $V_{XAP}(s)$ has two components with a component directly proportional to the iAP $V'_{mAP}(s)$ itself due to the leaking nanojunctional membrane plus a scaled negative first time derivative of the iAP due to the junctional membrane current. The second term is actually a voltage-divided version of the eAP at the junctional cleft (i.e., $V_{seal}(s)$). Therefore, the $V_{XAP}(s)$ at Point X during membrane poration comprises fractions of both the iAP and eAP. As the amplitude of the first term is substantially (two orders of magnitude) larger than that of the second, the overall $v_{XAP}(t)$ looks like a scaled version of the iAP (Dipalo et al. 2017; Xie et al. 2012; Lin et al. 2014; Robinson et al. 2012).

The recorded signal $V_{in}(s)$ is still determined by Eq. (9.2), further incorporating both an amplitude attenuation and a phase distortion. In the literature, this type of recorded signal is termed as "intracellular-like AP" (Dipalo et al. 2017), which is a reasonable description, as the recording is neither the full-stroke iAP, nor simply a scaled version of it, though their waveforms may seem very similar. The difference in the amplitudes of the $v_{XAP}(t)$ before and after nanojunctional membrane poration (Eq. (9.4) vs. (9.7)) is a direct result of the difference between the nanojunctional membrane current $i_{njm}(t)$ (in pA) and the leaking current $i'_{njm}(t)$ (in nA), which reflects a substantially enhanced recording after membrane poration (Dipalo et al. 2017; Xie et al. 2012; Robinson et al. 2012).

9.3   Recording by Multiple Nanoprotrusion Electrodes on the Same Planar Microelectrode

Our discussions above can be directly extended to multiple nanoprotrusion electrodes fabricated on the same planar microelectrode as illustrated in Fig. 9.1 (Dipalo et al. 2017; Xie et al. 2012; Robinson et al. 2012). The equivalent electrical circuits with n nanoprotrusion electrodes are shown in Figs. 9.4 and 9.5. If the surface of the planar microelectrode is passivated (Xie et al. 2012; Robinson et al. 2012), these circuits are the same as those with only one nanoprotrusion electrode in Figs. 9.2 and 9.3, except for a scaling factor n to the parameters in the nanojunction (Guo 2019) (see Exercise 9.3).

9.3.1   Extracellular Recording

For extracellular recording (Fig. 9.4), with parameter substitutions in Eqs. (9.3) and (9.4), we have

$$
\begin{aligned}
V_{Xsub}(s) &\approx \left[\beta_{njm}R_{njseal} + \left(\beta_{jm} + n\beta_{njm}\right)R_{jseal}\right]C_m \cdot sV_{msub}(s) \\
&\approx \left[\beta_{njm}R_{njseal} + \left(\beta_{jm} + \beta_{njm}\right)R_{jseal}\right]C_m \cdot sV_{msub}(s)
\end{aligned}
\tag{9.8}
$$

$$
\begin{aligned}
V_{XAP}(s) &= -\left[\beta_{njm}R_{njseal} + \left(\beta_{jm} + n\beta_{njm}\right)R_{jseal}\right]C_m \cdot sV_{mAP}(s) \\
&\approx -\left[\beta_{njm}R_{njseal} + \left(\beta_{jm} + \beta_{njm}\right)R_{jseal}\right]C_m \cdot sV_{mAP}(s)
\end{aligned}
\tag{9.9}
$$

$V_X(s)$ approximates to that in the single nanoprotrusion electrode case. This is because (1) the voltage across the nanojunctional resistance stays the same as the factor n's in $nI_{njm}(s)$ or $-nI_{njm}(s)$ and $\frac{R_{njseal}}{n}$ cancel with each other and (2) the additional current $(n-1)I_{njm}(s)$ or $-(n-1)I_{njm}(s)$ (still very tiny due to small β_{njm}) flowing through R_{jseal} only produces a tiny voltage increase in $V_{seal}(s)$.

Regarding the recording, with parameter substitution in Eq. (9.2), we have

$$
V_{in}(s) = \frac{Z_{in}}{\widetilde{Z}''_e + Z_{in}} V_X(s)
\tag{9.10}
$$

where $\widetilde{Z}''_e = \frac{Z_e}{n} + \frac{R_{njseal}}{n} + R_{jseal}$. The recording $v_{in}(t)$ becomes substantially larger, as a result of reduction of the equivalent *in-situ* impedance Z''_e of the nanoprotrusion electrode by approximately n times.

With an unpassivated substrate microelectrode surface (Dipalo et al. 2017), the effects are the same as the general case discussed in Chap. 6, Sect. 6.2. That is, the EDL capacitance C'_{dl} of the cell-covered electrode surface functions as a regular

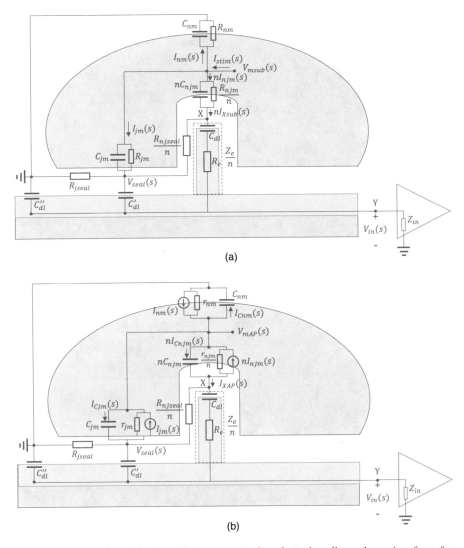

Fig. 9.4 Abstracted models of multiple nanoprotrusion electrode–cell membrane interfaces for extracellular recording. (**a**) During subthreshold depolarization. (**b**) During AP. Note that the n nanoprotrusion electrodes are illustrated as an equivalent nanoprotrusion electrode with proper parameter adjustments. Adapted with permission from (Guo 2019). Copyright © 2019 IOP Publishing Ltd

planar recording electrode to sense the eFP $v_{seal}(t)$ across R_{jseal}, which is superimposed onto the recording $v_{in}(t)$. It is noted that the eFP $v_{seal}(t)$ is much smaller than the eFP $v_X(t)$ at the nanoprotrusion electrode's tip, as R_{jseal} is more than three orders of magnitude smaller than R_{njseal}. Meanwhile, the EDL capacitance C''_{dl} of the electrode surface not covered by the cell acts to shunt the amplifier's input impedance directly, attenuating the recording $v_{in}(t)$.

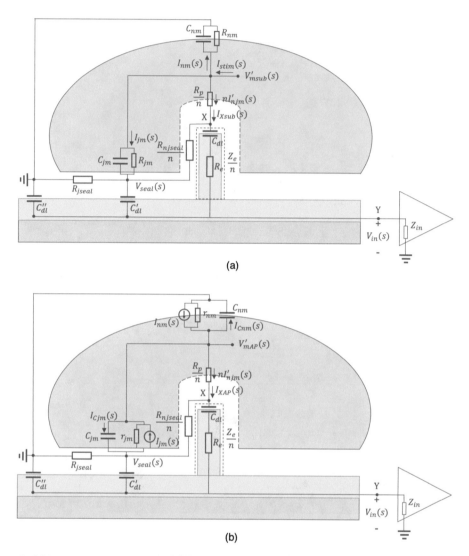

Fig. 9.5 Abstracted models of multiple nanoprotrusion electrode–cell membrane interfaces for intracellular-like recording after membrane poration. (**a**) During subthreshold depolarization. (**b**) During AP. Note that the n nanoprotrusion electrodes are illustrated as an equivalent nanoprotrusion electrode with proper parameter adjustments. Adapted with permission from (Guo 2019). Copyright © 2019 IOP Publishing Ltd

9.3.2 Recording After Membrane Poration

After membrane poration (Fig. 9.5), when the surface of the planar microelectrode is passivated (Xie et al. 2012; Robinson et al. 2012), with parameter substitutions in Eqs. (9.6) and (9.7), we have

$$V_{Xsub}(s) = \frac{nR_{jseal} + R_{njseal}}{nR_{jseal} + R_p + R_{njseal}} V'_{msub}(s) \approx \frac{R_{jseal} + R_{njseal}}{R_{jseal} + R_p + R_{njseal}} V'_{msub}(s) \quad (9.11)$$

$$V_{XAP}(s) \approx \frac{R_{njseal}}{R_p + R_{njseal}} V'_{mAP}(s) - \frac{R_p}{R_p + R_{njseal}} R_{jseal} C_{jm} \cdot s V'_{mAP}(s) \quad (9.12)$$

$V_{Xsub}(s)$ approximates to that in the single nanoprotrusion electrode case, whereas $V_{XAP}(s)$ stays the same. $V_{in}(s)$ is still determined by Eq. (9.10) and is substantially enhanced due to reduction of the *in-situ* electrode impedance Z''_e by approximately n times.

With an unpassivated substrate microelectrode surface (Dipalo et al. 2017), the EDL capacitance C'_{dl} of the cell-covered electrode surface functions as a regular planar recording electrode to sense the eFP $v_{seal}(t)$ across R_{jseal}, which is superimposed onto the recording $v_{in}(t)$. However, $v_{seal}(t)$ now has a small fraction of $v'_m(t)$ due to the leaking currents $ni'_{njm}(t)$ from the nanojunctions, which does not affect $v_{in}(t)$ much as R_{jseal} is more than three orders of magnitude smaller than R_{njseal}.

9.4 Conclusion

In this chapter, we use equivalent electrical circuit models of the neuron-nanoelectrode interfaces to derive closed-form analytical relationships between the iAP $v_m(t)$ and eFP $v_X(t)$ during subthreshold depolarization and suprathreshold AP, respectively. Such closed-form solutions offer a clear and complete understanding on the recording mechanisms, nature of signals, and interplays between key interface parameters.

This general theoretical framework is not configuration-specific and can be adapted to a variety of extracellular and intracellular recording situations, including using the vertical nanowire electrode arrays in the "Faradaic" regime (Robinson et al. 2012), the gold plasmonic nanocylindrical electrode arrays with optoporation only at the electrode tips (Dipalo et al. 2017), and the gold mushroom-shaped microelectrode (Hai et al. 2010a, b; Hai and Spira 2012). These findings have broad implications to advance the theory and practice of nano neurotechnologies, including offering critical insights to the proper design, characterization, and usage of this class of nanoelectrodes.

9.5   Summary

- During extracellular recording with a single nanoprotrusion electrode, $V_{Xsub}(s)$ is proportional to the first time derivative of $V_{msub}(s)$, and $V_{XAP}(s)$ is proportional to the negative first time derivative of the iAP $V_{mAP}(s)$.
- After membrane poration, $V_{Xsub}(s)$ is simply a fraction of $V'_{msub}(s)$ according to a voltage-divider circuit. It is discovered that no current is flowing across the junctional membrane, i.e., $I_{jm}(s) = 0$. The eAP $V_{XAP}(s)$ has two components with a component directly proportional to the iAP $V'_{mAP}(s)$ itself due to the leaking nanojunctional membrane plus a scaled negative first time derivative of the iAP due to the junctional membrane current. This second term is actually a voltage-divided version of the eAP at the junctional cleft (i.e. $V_{seal}(s)$).
- During extracellular recording by multiple nanoprotrusion electrodes on the same passivated planar microelectrode, the signal $V_X(s)$ approximates to that in the single nanoprotrusion electrode case. However, the recording $v_{in}(t)$ becomes substantially larger, as a result of reduction of the equivalent in-situ impedance Z''_e of the nanoprotrusion electrode by approximately n times.
- During extracellular recording by multiple nanoprotrusion electrodes with an unpassivated substrate microelectrode surface, the EDL capacitance C'_{dl} of the cell-covered electrode surface functions as a regular planar recording electrode to sense the eFP $v_{seal}(t)$ across R_{jseal}, which is superimposed onto the recording $v_{in}(t)$. Meanwhile, the EDL capacitance C''_{dl} of the electrode surface not covered by the cell acts to shunt the amplifier's input impedance directly, attenuating the recording $v_{in}(t)$.
- During recording by multiple nanoprotrusion electrodes after membrane poration, when the surface of the planar microelectrode is passivated, $V_{Xsub}(s)$ approximates to that in the single nanoprotrusion electrode case, whereas $V_{XAP}(s)$ stays the same. $V_{in}(s)$ is substantially enhanced due to reduction of the in situ electrode impedance Z''_e by approximately n times. When the planar microelectrode is unpassivated, the EDL capacitance C'_{dl} of the cell-covered electrode surface functions as a regular planar recording electrode to sense the eFP $v_{seal}(t)$ across R_{jseal}, which is superimposed onto the recording $v_{in}(t)$. However, $v_{seal}(t)$ now has a small fraction of $v'_m(t)$ due to the leaking currents $ni'_{njm}(t)$ from the nanojunctions, which does not affect $v_{in}(t)$ much as R_{jseal} is more than three orders of magnitude smaller than R_{njseal}.

Note: This chapter was adapted from (Guo 2019) with permission.

Exercises

9.1 Without considering the electrode conductor, prove that the nanojunctional seal exerts a minimum effect on $I_{njm}(s)$ in Fig. 9.2a, so that the entire cell membrane can be considered to have a uniform transmembrane current density and $I_{njm}(s)$ can be approximated as $\beta_{njm}I_{stim}(s)$.

9.2 A nanoprotrusion electrode is able to record EPSP and IPSP extracellularly thanks to the ultra-tight nanojunctional seal. How do the extracellularly recorded EPSP and IPSP look like?

9.3 Referring to Fig. 9.1, a neuron is sitting on top of an array of n nanoprotrusion electrodes densely packed on a passivated planar microelectrode. The equivalent electrical circuit for the extracellular subthreshold recording is provided in (a) below. Prove that the circuit in (a) is equivalent to the circuit in (b) which is the same circuit as in Fig. 9.4a.

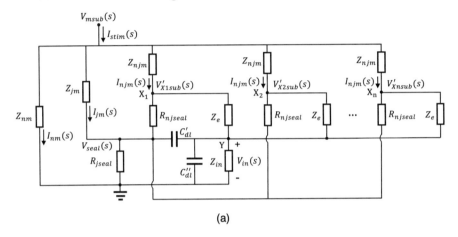

(a)

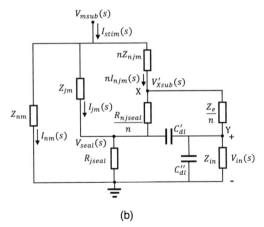

(b)

9.4 Explain:

(a) Whether multiple nanoprotrusion electrodes on the same planar microelectrode should be used as opposed to a single nanoprotrusion electrode and why?

(b) Whether the underlying planar microelectrode should be passivated and why?

(c) Why to record after membrane poration?

References

Dipalo M, Amin H, Lovato L, Moia F, Caprettini V, Messina GC, Tantussi F, Berdondini L, De Angelis F (2017) Intracellular and Extracellular Recording of Spontaneous Action Potentials in Mammalian Neurons and Cardiac Cells with 3D Plasmonic Nanoelectrodes. Nano Lett 17 (6):3932-3939. doi:https://doi.org/10.1021/acs.nanolett.7b01523

Gold C, Henze DA, Koch C, Buzsaki G (2006) On the origin of the extracellular action potential waveform: A modeling study. Journal of neurophysiology 95 (5):3113-3128. doi:https://doi.org/10.1152/jn.00979.2005

Grattarola M, Martinoia S (1993) Modeling the neuron-microtransducer junction: from extracellular to patch recording. IEEE transactions on bio-medical engineering 40 (1):35-41. doi:https://doi.org/10.1109/10.204769

Guo L (2019) On neural recording using nanoprotrusion electrodes. J Neural Eng 17 (1):016017

Hai A, Shappir J, Spira ME (2010a) In-cell recordings by extracellular microelectrodes. Nature methods 7 (3):200-U250. doi:https://doi.org/10.1038/Nmeth.1420

Hai A, Shappir J, Spira ME (2010b) Long-Term, Multisite, Parallel, In-Cell Recording and Stimulation by an Array of Extracellular Microelectrodes. Journal of neurophysiology 104 (1):559-568. doi:https://doi.org/10.1152/jn.00265.2010

Hai A, Spira ME (2012) On-chip electroporation, membrane repair dynamics and transient in-cell recordings by arrays of gold mushroom-shaped microelectrodes. Lab on a Chip 12 (16):2865-2873

Lewandowska MK, Bakkum DJ, Rompani SB, Hierlemann A (2015) Recording large extracellular spikes in microchannels along many axonal sites from individual neurons. PloS one 10 (3):e0118514. doi:https://doi.org/10.1371/journal.pone.0118514

Lin ZLC, Xie C, Osakada Y, Cui Y, Cui BX (2014) Iridium oxide nanotube electrodes for sensitive and prolonged intracellular measurement of action potentials. Nat Commun 5. doi: https://doi.org/10.1038/ncomms4206

Nelson MJ, Pouget P, Nilsen EA, Patten CD, Schall JD (2008) Review of signal distortion through metal microelectrode recording circuits and filters. J Neurosci Meth 169 (1):141-157. doi:https://doi.org/10.1016/j.jneumeth.2007.12.010

Oppenheim A, Willsky A (1996) Signals and Systems. 2nd edn. Pearson,

Robinson JT, Jorgolli M, Shalek AK, Yoon MH, Gertner RS, Park H (2012) Vertical nanowire electrode arrays as a scalable platform for intracellular interfacing to neuronal circuits. Nature nanotechnology 7 (3):180-184

Tian BZ, Cohen-Karni T, Qing Q, Duan XJ, Xie P, Lieber CM (2010) Three-Dimensional, Flexible Nanoscale Field-Effect Transistors as Localized Bioprobes. Science 329 (5993):830-834. doi: https://doi.org/10.1126/science.1192033

Wang Y, Zhu H, Yang H, Argall AD, Luan L, Xie C, Guo L (2018) Nano functional neural interfaces. Nano Res. doi: https://doi.org/10.1007/s12274-018-2127-4

Xie C, Lin ZL, Hanson L, Cui Y, Cui BX (2012) Intracellular recording of action potentials by nanopillar electroporation. Nature nanotechnology 7 (3):185-190. doi:https://doi.org/10.1038/Nnano.2012.8

Chapter 10
Recording Using Tetrodes

When a single isotropic microelectrode (e.g., a sphere or column electrode) is concerned, the neuronal current source can only be determined to be within the sphere of its receptive field (see Chap. 11 for definition), i.e., no specific 3D coordinates of the current source can be resolved. As the radius of a microelectrode's receptive field can be larger than 100μm (Mechler et al. 2011; Delgado Ruz and Schultz 2014), this results in a poor spatial resolution of the neuronal source. When multiple microelectrodes are used as a group to concurrently record from the same neuron, these multiple recordings can simply be superimposed to cancel noises and yield an average recording with an improved SNR. Another way of using multiple microelectrodes is to improve the performance of spike sorting, comparing to simple use of individual electrodes. A still more advanced technique is to use a group of spatially distributed microelectrodes to solve the inverse problem of current source localization by taking advantage of the capability of microelectrode arrays (MEAs). To achieve this goal, the minimum electrode number is three for microelectrodes located on a large planar probe, which cuts the recording volume into two halves, to map a current source on the front side of the volume and is four for microelectrodes located on a small non-planar probe, which has an isotropic overlapping receptive field of the four electrodes, to map a current source in a full 3D volume. The interelectrode spacings need to be carefully controlled so that, on the one hand, all the electrodes can record the same current source concurrently, and, on the other hand, the spacings should be larger than the spatial resolution (see Chap. 11) of the electrode recording system. Such a non-planar microelectrode combination is called a *tetrode*. Figure 10.1 illustrates a typical tetrode arranged as a tetrahedron in a conical probe (such as the conical Thomas tetrodes).

A tetrode forms a typical electrode unit (see Chap. 11), by which a neuronal current source falling within its overlapping receptive field can be localized (Lee et al. 2013; Chelaru and Jog 2005; Mechler et al. 2011). However, because a tetrode's overlapping receptive field generally includes a number of neuronal sources due to the large radius of each electrode's receptive field, the tetrode itself is incapable of isolating and identifying spikes from an individual source, so that a

Fig. 10.1 Illustration of a typical tetrode arranged as a tetrahedron in a conical probe's tip

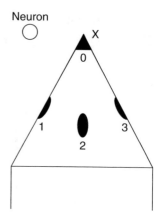

spike sorting and classification algorithm has to be used to cluster and identify the spikes. The inverse problem of source localization can then be solved after a spike from a neuron is identified in the four electrodes' concurrent recordings. As a result, the errors and inaccuracies of spike sorting algorithms add to the intrinsic errors and inaccuracies of the source localization approach using tetrodes. In this book, we won't go into the details of spike sorting and classification algorithms, but the readers can check some excellent reviews on this topic (Lewicki 1998; Wild et al. 2012; Rey et al. 2015). In the following, we will elaborate the mathematical principle of source localization using a tetrode under a monopole source model, starting with deriving an analytical solution. When an analytical solution does not exist due to complicated situations, the inverse problem can be reformulated as an optimization problem which aims to minimize a cost function.

10.1 Principle of Source Localization

10.1.1 Why the Neuronal Source Is Viewed as a Current Source?

When a neuron fires an AP, its perisomatic region emits a dominant AP current (i.e., the $-i_{AP}(t)$) that produces eFPs in its surrounding resistive extracellular space (see Chaps. 2 and 6). Although an equivalent potential exists on the outer perisomatic surface, this perisomatic region is better viewed as an equivalent current source. Different equivalent current source models have been developed to account for the morphological and electrical properties of the neuronal current source. In addition to its 3D location, the strength of this equivalent current source is also a key parameter to be determined.

10.1.2   Monopole Source Model

The dynamic electric field generated by a firing neuron in the extracellular space is spatiotemporally sampled as time-varying voltages (i.e., the electromotive forces to move charges) by multiple electrodes. At a certain moment, a set of spatial voltage samples of the electric field can be used to deduce the location of the current source when the properties of the source are known, including the source's shape and the spatially varying rule of its electric field. However, such information is unknown in almost all intracortical recordings where localization of the source is desired. To approach this inverse problem, reasonable assumptions on the source's spatial morphology and the spatially varying rule of its electric field have to be made with a proposal of an equivalent current source model, such as the monopole (Lee et al. 2007; Guo 2020b), dipole (Mechler et al. 2011; Mechler and Victor 2012), or line source model (Delgado Ruz and Schultz 2014), to constrain the inverse problem for a possible solution. The solution to this inverse problem thus fits the forward predictions of the equivalent current source model to the spatially sampled voltages. Each of these source models has its predicting power and limitations, depending on the actual type and morphology of the neuron to be mapped. Inaccuracy to the solution to the current source localization primarily stems from the mismatches between the equivalent current source model and the real current-emitting morphology of the neuron and between the assumed and actual spatially varying rules of its extracellular electric field.

With only four parameters (strength, and three spatial coordinates), the monopole equivalent current source model is the simplest model to account for the dominant perisomatic extracellular current source from a firing neuron. So, we use it here to elaborate the principle of source localization analytically from a tetrode's recordings.

As mentioned above, localization precision strongly depends on how well the assumed source model matches the characteristics of the actual source (Mechler et al. 2011). In some source localization studies using tetrodes, a dipole source model is used (Delgado Ruz and Schultz 2014; Mechler et al. 2011; Mechler and Victor 2012). Simulation results (Delgado Ruz and Schultz 2014) showed that when the electrode is very close to the perisomatic source, a dipole model is more accurate, however, when the electrode is relatively far away from the source, a monopole model can faithfully represent the perisomatic source. In the perisomatic region, the strongest transmembrane currents are found in the axon hillock/initial segment area. Thus, it is expected that the estimated source location/center will be somewhere in the vicinity of the soma–axon hillock–initial segment complex, rather than the geometric center of the soma. Given such an irregular source morphology (such as that of a pyramidal neuron) which is relatively large comparing to a microelectrode in a typical tetrode, when the tetrode is placed at a close distance to different parts of the neuron, variations of the eAP are recorded (Delgado Ruz and Schultz 2014). However, for a tetrode which is placed either far away from the source or close to a certain part of the source (for a very small tetrode), all four electrodes should record an eAP with a consistent waveform, where an additional minor copy of the eAP might superimpose onto the major copy causing a slight variation of the overall waveform. But in most cases, such a minor copy may not be detected above the noise floor of an electrode recording system, which has a typical root-mean-square (RMS) noise amplitude of 12–35μV for a tetrode channel (Mechler and Victor 2012).

For simplicity, in Fig. 10.1, we assume each electrode in the tetrode to have the same equivalent geometry and properties, and each recording channel has the same $\frac{Z'_{ain}}{Z_e+Z'_{ain}+R_s}$ according to Eq. (3.3). We ignore the physical sizes of the neuron and electrodes for now and will include them in consideration in Chap. 11 for advanced analyses. We model the neuronal current source as an equivalent monopole current source whose center corresponds to the averaged center location of a neuron's soma–axon hillock–initial segment complex. We refer to the source location as the location of this averaged center. We further assume a simple volume conductor model of the medium: the resistive volume conductor is homogeneous so that the extracellular perisomatic current $i_E(t)$ generates an extracellular electric field with isopotential spheres centering at the point current source according to

$$v_X(t) = \frac{i_E(t)}{4\pi\sigma r} \tag{10.1}$$

where σ is the conductivity of the medium and r is the distance the isopotential sphere from the center of the point current source.

10.1.3 Analytical Solution to the Inverse Problem

For a tetrode located at Point X (i.e., the location of Electrode #0 in Fig. 10.1) with a monopole neuronal current source in its overlapping receptive field, all of its four electrodes can have a recording $v_{ini}(t)$ when an AP is fired in the neuron. According to Eq. (3.3), we label the corresponding deconvoluted eAPs as $v_{X0}(t)$, $v_{X1}(t)$, $v_{X2}(t)$, and $v_{X3}(t)$, in reference to each electrode's number in the tetrode.

To successfully locate the soma, it is necessary to select the instant at which the currents are concentrated near the soma (Delgado Ruz and Schultz 2014). The major negative peak of an eAP corresponds to the largest inward Na^+ current during the depolarization phase (Eq. (6.2), also see Fig. 6.2) (Guo 2020a) and is primarily contributed by the perisomatic region, thus it should be used to localize the soma (Mechler et al. 2011). In contrast, the second positive peak corresponds to the largest K^+ outflow current during the repolarization phase and is not well confined to the perisomatic region, thus should not be used for localizing the soma (Lee et al. 2007).

According to Eq. (10.1), the amplitude (the eAP's amplitude is negative according to Eq. (6.5)) of the eAP is caused by the peak of the extracellular current; and the difference in the amplitude of each electrode's eAP is caused by their different distances from the current source. To approach this inverse problem, we write the eAP's amplitude as

$$v_{X\_peak} = k\frac{1}{r} \tag{10.2}$$

where $k = \frac{i_{E\_peak}}{4\pi\sigma}$. From the tetrode's four concurrent recordings on the same spike, we obtain v_{X0_peak}, v_{X1_peak}, v_{X2_peak}, and v_{X3_peak}. Thus, we have a set of four equations:

$$\begin{cases} v_{X0\_peak} = k\dfrac{1}{r_0} \\[2mm] v_{X1\_peak} = k\dfrac{1}{r_1} \\[2mm] v_{X2\_peak} = k\dfrac{1}{r_2} \\[2mm] v_{X3\_peak} = k\dfrac{1}{r_3} \end{cases} \tag{10.3}$$

We set the origin of our coordinate system at the center location of Electrode #0 (E0), and the relative 3D coordinates of the other three electrodes are known according to the tetrode's design. We have the coordinates of the four electrodes as E0: $(0,0,0)$, E1: (x_1, y_1, z_1), E2: (x_2, y_2, z_2), and E3: (x_3, y_3, z_3). We label the location of the current source as S: (x_s, y_s, z_s). Thus,

$$\begin{cases} r_0 = \sqrt{x_s^2 + y_s^2 + z_s^2} \\[2mm] r_1 = \sqrt{(x_s - x_1)^2 + (y_s - y_1)^2 + (z_s - z_1)^2} \\[2mm] r_2 = \sqrt{(x_s - x_2)^2 + (y_s - y_2)^2 + (z_s - z_2)^2} \\[2mm] r_3 = \sqrt{(x_s - x_3)^2 + (y_s - y_3)^2 + (z_s - z_3)^2} \end{cases} \tag{10.4}$$

Substituting these r's into Eq. (10.3), we have four independent variables (x_s, y_s, z_s, and k) and four independent equations. Given the physical constraints, a unique analytical solution can be derived (Lee et al. 2007; Guo 2020b), though the mathematical procedure is very complicated. From this solution, k is also solved, so that i_{E_peak}, the source strength (Lee et al. 2007), can also be derived.

If all four electrodes are on the same plane, we would only have three independent equations in Eq. (10.3), and the solution to the four-variable problem cannot be uniquely determined. However, in the case of a large multi-site planar neural probe when an additional condition to this inverse problem is added by constraining the neuronal current source to be on one side of the probe, three electrodes in the plane of the probe can locate the source by sorting out a satisfying solution under this additional constraint from an infinite number of solutions to the three equations.

10.1.4 Whether Deconvolution is Needed?

According to Eqs. (10.1) and (10.2), the eAPs and their amplitudes recorded by the four electrodes of a tetrode differ only by a constant factor related to the ratio of their $\frac{1}{r}$'s, so that the eAPs' Fourier transforms as functions of the frequency variable $j\omega$ also differ by the same scaling factors. We write $v_{Xi}(t) = \alpha_{i0}v_{X0}(t)$, where $i = 1, 2, 3$ and $\alpha_{i0} = \frac{r_0}{r_i}$, thus, we have $V_{Xi}(j\omega) = \alpha_{i0}V_{X0}(j\omega)$. Further according to Eq. (3.3), $V_{ini}(j\omega) = V_{Xi}(j\omega)Z(j\omega)$, where $Z(j\omega) = \frac{Z'_{ain}}{Z_e + Z'_{ain} + R_s}$ for each recording channel of the four electrodes in a tetrode, thus, $V_{ini}(j\omega) = \alpha_{i0}V_{X0}(j\omega)Z(j\omega) = \alpha_{i0}V_{in0}(j\omega)$, which gives us the time-domain equation of $v_{ini}(t) = \alpha_{i0}v_{in0}(t)$. Therefore, the direct recordings $v_{ini}(t)$ of the four electrodes also differ by the same set of constant factors, and we can replace Eq. (10.3) by

$$
\begin{cases}
v_{in0\_peak} = k'\dfrac{1}{r_0} \\[2mm]
v_{in1\_peak} = k'\dfrac{1}{r_1} \\[2mm]
v_{in2\_peak} = k'\dfrac{1}{r_2} \\[2mm]
v_{in3\_peak} = k'\dfrac{1}{r_3}
\end{cases}
\tag{10.5}
$$

where $k' = \frac{[i_E(t)*z(t)]_{peak}}{4\pi\sigma}$ and "*" denote the convolution operation. Combining Eqs. (10.4) and (10.5), x_s, y_s, z_s, and k' can be solved using the direct recordings $v_{ini}(t)$'s *without the need of performing deconvolution*. But if we need to recover the source strength i_{E_peak}, we still need to perform deconvolution on at least one direct recording $v_{ini}(t)$ to recover the eAP $v_{Xi}(t)$ and then derive i_{E_peak} using Eq. (10.2).

10.2 When a Real Analytical Solution Does Not Exist

A real analytical solution to Eq. (10.3) or (10.5) may not exist for a few reasons: (1) the real morphological AP current distribution of the neuron substantially deviates from the assumed radial distribution of an equivalent monopole somatic current source, (2) the spatially varying rule of generated extracellular electric field does not obey the $\frac{1}{r}$ characteristics as shown in Eq. (10.1), (3) the inhomogeneity of the volume conductor distorts the extracellular potentials, (4) the presence of unknown noises cause errors, and (5) the $Z(j\omega)$ of each electrode recording channel is not consistent. To circumvent these situations, this ill-posed inverse problem can be reformulated as an optimization problem by minimizing a cost function, e.g., the mismatch between the measured and calculated eAPs, for a reasonable approximate solution (Lee et al. 2007). And, indeed, minimization of an objective function is often used to derive the source location for other more complicated equivalent current source models, such as the dipole model (Mechler and Victor 2012), when an analytical solution is hard or impossible to find. However, it should be noted that simply minimizing the total error can lead to severe biases in localization, because the minimization will choose a position for the source at which errors arising from the above causes happen to cancel each other (Mechler et al. 2011).

10.3 Stepping Tetrode

An effective way of using the tetrode is to vertically step it, e.g., at 10μm increment, using a computer-controlled precision microdrive (Fig. 10.2) (Mechler et al. 2011). This stepping tetrode approach allows the source mapping of a vertical cortical column of a radius equal to that of the tetrode's receptive field, e.g., 100μm (Mechler and Victor 2012). The neuronal source localization has a linear resolution of 50μm in each of the x, y, and z directions. On average, a tetrode isolated six single neurons along a 90-μm-long track (the most was 13), and it was estimated that the isolated fraction of all neurons present within the recording volume was only about 1% or smaller (Mechler et al. 2011). Again, spike sorting and classification are required before source localizations can be performed, and the localized sources are only

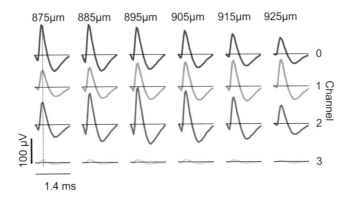

Fig. 10.2 Typical intracortical recordings from a single unit by a stepped tetrode. The spike waveforms (average of 900–1,400 spikes/step) registered by the four channels of the tetrode are shown in separate columns for each of the six equidistant recording positions. Note, upward deflections represent negative eAPs. Reproduced with permission from (Mechler et al. 2011). Copyright © 2011 the American Physiological Society

estimations prone to errors and inaccuracies. Furthermore, multiple stepping tetrodes can be inserted in parallel in a cortical region to form a "pseudo" 3D MEA. However, the lateral spacing between these tetrodes cannot be made small enough and maintained consistently during stepping, making it impossible to completely and exactly identify and localize all the active eAP current sources within the recording volume. Lastly, the multiple stepping tetrode approach is inapplicable to chronic recording in awake animals, with tethering to the microdrives and reversible stepping and withdrawing the tetrodes in the recording paths. These problems are exactly what Chap. 11 aims to address through an integrated 3D ultra-density MEA.

10.4   Limitations of Tetrodes

While tetrodes are a valuable tool for neuroelectrophysiology, based on the above discussions, the major functional limitations of the tetrode approach on source localization include: (1) reliance on spike sorting and classification algorithms to identify the common spikes from the same neuron, (2) proneness to errors and inaccuracies or even lack of an analytical solution altogether for various reasons, and (3) limited source resolving capability in the recording volume and thus incapable of reconstructing the spatial configuration of local neuronal microcircuits.

10.5   Summary

- Using a tetrode, the inverse problem of source localization can be solved after a spike from a neuron is identified in the four electrodes' concurrent recordings using a spike sorting algorithm.
- The perisomatic region of a neuron is typically viewed as an equivalent current source. In addition to its 3D location, its strength is also a key parameter to be determined.
- To approach the inverse problem of source localization, reasonable assumptions on the source's spatial morphology and the spatially varying rule of its electric field have to be made with a proposal of an equivalent current source model, such as the monopole model, to constrain the inverse problem for a possible solution. Inaccuracy to the solution primarily stems from the mismatches between the equivalent current source model and the real current-emitting morphology of the neuron and between the assumed and actual spatially varying rules of its extracellular electric field.
- With only four parameters (strength, and three spatial coordinates), the monopole equivalent current source model is the simplest model to account for the dominant perisomatic extracellular current source from a firing neuron. When the electrode is relatively far away from the source, a monopole model can faithfully represent the perisomatic source.

- The major negative peak of an eAP corresponds to the largest inward Na$^+$ current during the depolarization phase and is primarily contributed by the perisomatic region, thus it should be used to localize the soma.
- The location and strength of the neuronal current source are solved using Eqs. (10.3) and (10.4), which have four independent equations for four independent variables (x_s, y_s, z_s, and k).
- To localize the neuronal current source, deconvolution is not needed to recover the eAPs from the direct recordings, but it is needed to unravel the source strength.
- For various reasons, when a real analytical solution does not exist, the ill-posed inverse problem can be reformulated as an optimization problem by minimizing a cost function, e.g., the mismatch between the measured and calculated eAPs, for a reasonable approximate solution.
- A tetrode can also be stepped stepwise to resolve the neuronal sources along a track, and multiple stepping tetrodes can be used together to sample along multiple tracks.

Exercises

10.1 What are tetrodes typically used for?

10.2 Answer:

(a) Why is the neuronal source viewed as a current source?

(b) What are the available models for the neuronal current source? What are their pros and cons?

(c) In what situations is the monopole equivalent current source model a reasonable representation of the actual neuronal current source?

10.3 For source localization using a tetrode,

(a) Describe the principle of source localization.

(b) Is deconvolution necessary to solving the inverse problem?

(c) How to approach the problem when a real analytical solution does not exist? What are the potential pitfalls?

10.4 Answer:

(a) What are the advantages and disadvantages of stepping a tetrode?

(b) What are the limitations of the tetrode approach in general?

10.5 Referring to Fig. 10.2, given our prior knowledge from Chaps. 6 and 9 that without an ultra-tight membrane-electrode seal, subthreshold depolarizations cannot be detected above the noise level of a conventional microelectrode recording system, why are the first positive peaks of the eAPs shown up? (Note that the upward deflections represent negative eAPs.)

References

Chelaru MI, Jog MS (2005) Spike source localization with tetrodes. J Neurosci Methods 142 (2):305-315. doi:https://doi.org/10.1016/j.jneumeth.2004.09.004

Delgado Ruz I, Schultz SR (2014) Localising and classifying neurons from high density MEA recordings. J Neurosci Methods 233:115-128. doi:https://doi.org/10.1016/j.jneumeth.2014.05.037

Guo L (2020a) Perspectives on electrical neural recording: a revisit to the fundamental concepts. J Neural Eng 17 (1):013001. doi:10.1088/1741-2552/ab702f

Guo L (2020b) Principles of functional neural mapping using an intracortical ultra-density micro-electrode array (ultra-density MEA). J Neural Eng 17. doi:https://doi.org/10.1088/1741-2552/ab8fc5

Lee CW, Dang H, Nenadic Z (2007) An efficient algorithm for current source localization with tetrodes. Conference proceedings : Annual International Conference of the IEEE Engineering in Medicine and Biology Society IEEE Engineering in Medicine and Biology Society Conference 2007:1282-1285. doi:https://doi.org/10.1109/IEMBS.2007.4352531

Lee CW, Szymanska AA, Ikegaya Y, Nenadic Z (2013) The accuracy and precision of signal source localization with tetrodes. Conference proceedings : Annual International Conference of the IEEE Engineering in Medicine and Biology Society IEEE Engineering in Medicine and Biology Society Conference 2013:531-534. doi:https://doi.org/10.1109/EMBC.2013.6609554

Lewicki MS (1998) A review of methods for spike sorting: the detection and classification of neural action potentials. Network: Computation in Neural Systems 9 (4):R53-R78

Mechler F, Victor JD (2012) Dipole characterization of single neurons from their extracellular action potentials. Journal of computational neuroscience 32 (1):73-100

Mechler F, Victor JD, Ohiorhenuan I, Schmid AM, Hu Q (2011) Three-dimensional localization of neurons in cortical tetrode recordings. Journal of neurophysiology 106 (2):828-848. doi:https://doi.org/10.1152/jn.00515.2010

Rey HG, Pedreira C, Quiroga RQ (2015) Past, present and future of spike sorting techniques. Brain research bulletin 119:106-117

Wild J, Prekopcsak Z, Sieger T, Novak D, Jech R (2012) Performance comparison of extracellular spike sorting algorithms for single-channel recordings. J Neurosci Meth 203 (2):369-376

Chapter 11
Intracortical Functional Neural Mapping Using an Integrated 3D Ultra-Density MEA

Intracortical extracellular electrical neural recording using solid-state microelectrodes is a prevalent approach in addressing neurophysiological queries and implementing brain–computer interfacing systems. On the cellular scale, the neocortex is anatomically organized into vertical layers and functions in perpendicular microcolumns. How this microarchitecture determines behavioral cognitive functions is a subject of sustained interest in the neuroscience field. Addressing this question demands tools, on the one hand, with vertical access to all cortical layers at a cellular spatial resolution and an AP scale (i.e., millisecond) temporal resolution, and on the other hand, covering a relatively large lateral recording area. Functional recording/imaging at the neuronal circuit level aims to identify the association (and hopefully the causal relationship) between a system-level function and the signal flow pattern of neuronal computation within a neuronal microcircuit.

To decipher how a neuronal microcircuit functions at the network level with cellular or even subcellular resolution, the neuroscience community calls for clear coupled images on both the network's topology and its spatiotemporally evolving neural activities. While the relative conservative and stationary topology of a small to medium microcircuit can be imaged at sufficient resolution using optical approaches, e.g., two-photon laser scanning microscopy, the corresponding in situ functional mapping in a freely moving subject has been lagging behind due to significant technological barriers. Although optical functional cellular imaging techniques, such as imaging using voltage-sensitive dyes and genetically encoded voltage-sensitive fluorescent proteins (Wang et al. 2018; Rivnay et al. 2017; Peterka et al. 2011; Knopfel et al. 2006), have existed for a few decades, their inherent limitations prevent their use either in chronic in vivo applications or with sufficient SNR and/or temporal resolution. Motivated by the much better chronic reliability and performance of implantable MEAs, in more recent years, a variety of ultra-high-density MEAs have been created at an unprecedented electrode packing density (Scholvin et al. 2016; Jun et al. 2017; Du et al. 2011; Rios et al. 2016; Wei et al. 2018; Blanche et al. 2005; Luan et al. 2017), thanks to advances in micro/nanofabrication technology. Then, two questions arise with regard to (1) what

L. Guo, *Principles of Electrical Neural Interfacing*,
https://doi.org/10.1007/978-3-030-77677-0_11

electrode density is needed to sufficiently assess the spatial evolution of extracellular electrical signals manifested by a cortical neuronal microcircuit, i.e., to reach to the extracellular spatial resolution of neuronal information processing, and (2) how such an MEA should be designed and used. Amid the prevalent but rather vague advocations for "spatial oversampling" within the neural interfaces community, two recent theoretical works unraveled the ultimate capability and limitation of neural MEA technology in functional mapping of a cortical neuronal microcircuit (Kleinfeld et al. 2019; Guo 2020b), resulting in insightful conclusions to both guide the proper design of ultra-high-density MEAs and inform their proper usage.

Building upon these works and our discussions in Chap. 10, this chapter thus rationally scales up the tetrode concept to a fully integrated large-scale MEA at the ultimate electrode density for complete spatiotemporal resolution of all single-unit sources in the intracortical recording volume, with the purpose of reconstructing the 3D functional microarchitecture of the neuronal microcircuit. In this chapter, we approach this design and use problem of ultra-high-density MEAs for functional intracortical neuronal circuit mapping from a signal analysis perspective. Starting with quantitative derivations of key basic concepts, we define the concept of ultra-density MEA in the context for fully resolving the extracellular perisomatic AP (epsAP) current sources within its recording volume. Then, we elaborate the principle of using such an ultra-density MEA for functional neural mapping, followed by proposing a recursive approach to completely resolve all epsAP sources from the ultra-density MEA's recordings. Last, we discuss the limitations and implications of the ultra-density MEA concept.

11.1 Basic Concepts of a Single Electrode

In the context of intracortical extracellular recording, for simplicity, we consider an ideal spherical electrode and a monopole neuronal current source to derive the analyses. A practically implementable square-column microelectrode (see Fig. 11.2b) can directly adopt these analytical results under reasonable approximations.

11.1.1 Amplitude Resolution

The signal amplitude resolution of an electrode depends on both the system noise (δ) of the recording circuit and the impedances of the electrode (Z_e) and the amplifier input terminal (Z_{in}). As illustrated in Fig. 11.1 (Guo 2020b, a), the recording $v_{in}(t)$ appearing across the input terminals of a differential amplifier is related to the eFP $v_X(t)$ at the outer edge of the electrode's EDL, which is labeled as Point X, by

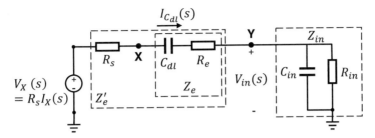

Fig. 11.1 The equivalent electrical circuit formed between the recording electrode and amplifier input. The eAP current propagating to the edge (Point X) of the EDL of the recording electrode is represented as a current source $I_X(s)$. Reproduced with permission from (Guo 2020b). Copyright © 2020 IOP Publishing Ltd

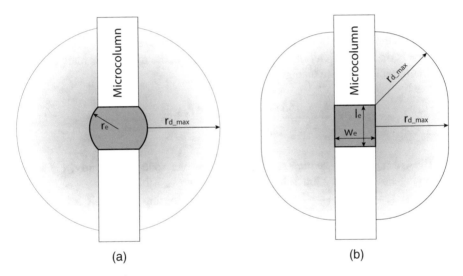

Fig. 11.2 Cross-sectional illustration of the receptive field of an electrode. (**a**) An ideal spherical electrode of radius r_e fabricated on a microcolumn. (**b**) A square-column electrode with $w_e = l_e = \sqrt{2}r_e$. w_e: side width, and l_e: length. Adapted with permission from (Guo 2020b). Copyright © 2020 IOP Publishing Ltd

$$V_{in}(s) = \frac{Z_{in}}{Z_e + R_s + Z_{in}} V_X(s) \qquad (11.1)$$

using frequency-domain representation. If we set the *minimum recordable signal amplitude* v_{in_min} by the recording circuit to be the maximum amplitude δ_{peak} of its input-referred noise, the *minimum detectable eFP* v_{X_min} is then determined as

$$v_{X\_min} = \left|1 + \frac{Z'_e}{Z_{in}}\right| \delta_{peak} = \Delta v_{X\_min} \qquad (11.2)$$

This v_{X_min} is also the *minimum potential change* Δv_{X_min} that can be detected by the electrode and its recording circuit. With the input impedance Z_{in} of the amplifier fixed (generally, it is recommended that $|Z_{in}| > 10|Z'_e|$ at least), a smaller electrode with a higher recording impedance Z_e thus has an unfavorable, larger Δv_{X_min}.

11.1.2 Spatial Resolution

Further considering that the eFP $v_X(t)$ is related to the distance r by which Point X is away from the location of the monopole current source $i_s(t)$ as $v_X(t) = R_s i_s(t) = \frac{i_s(t)}{4\pi\sigma r}$, where σ is the conductivity of the electrolyte solution, the Δv_{X_min} is mapped to the minimum radial distance change as

$$\Delta r_{min} = -\frac{r^2 \Delta v_{X\_min}}{r\Delta v_{X\_min} + \frac{i_{s\_peak}}{4\pi\sigma}} \Rightarrow \frac{\Delta r_{min}}{r} = -\frac{1}{1 + \frac{v_{X\_peak}}{\Delta v_{X\_min}}} \qquad (11.3)$$

Thus, Δr_{min} is the *spatial resolution* of the electrode. Apparently, (1) if the observing Point X has a much higher potential v_{X_peak} than the Δv_{X_min}, so that $\frac{v_{X\_peak}}{\Delta v_{X\_min}} \gg 1$, $\frac{\Delta r_{min}}{r} \approx -\frac{\Delta v_{X\_min}}{v_{X\_peak}}$; (2) if $\frac{v_{X\_peak}}{\Delta v_{X\_min}} \ll 1$, recording at Point X is practically meaningless, as no signal can be detected above the noise floor; and (3) if $\frac{v_{X\_peak}}{\Delta v_{X\_min}} = 1$, where the Point X is at the detectable edge ($r = r_{d\_max}$) of the *emission field* of the current source i_s, the spatial resolution is the poorest, with $\Delta r_{min} = \frac{r_{d\_max}}{2}$ (Mechler et al. 2011) (note, $r_{max} = r_{d\_max} + \Delta r_{min} = \frac{3}{2} r_{d\_max}$ is the radius of the monopole current source's emission field). From Eq. (11.3), we can see that, to have a good spatial resolution, i.e., a small Δr_{min}, the potential v_{X_peak} at the oberservation Point X needs to be much higher than the minimum detectable voltage Δv_{X_min}. The larger the current source i_{s_peak} and the closer to it the observation Point X, the larger v_{X_peak} is. Thus, a good spatial resolution is achieved at the vicinity of a large current source (Mechler et al. 2011), e.g., the perisomatic region of a neuron as opposed to its axon or dendrites, though their iAPs may have comparable amplitudes. Additionally, other conditions being the same, a smaller electrode has a poorer spatial resolution, i.e., a larger Δr_{min}. Spatial resolution of the electrode is important in determining the precision of 3D source localization. Thus, it is less precise to resolve axonal current sources than somatic ones.

11.1.3   Receptive Field

For a given monopole current source $i_s(t)$ with a spherical emission field defined of a radius r_{max}, where $v_{X\_peak} = 0$, the *receptive field* of an electrode is the volume enclosed by a surface with a distance of $r_{d\_max} = \frac{2}{3}r_{max}$ (see Sect. 11.1.2) from the electrode's surface. Usually, r_{max} is unknown, so we derive r_{d_max} from

$$r_{d\_max} = \frac{i_{s\_peak}}{4\pi\sigma\Delta v_{X\_min}} \tag{11.4}$$

Accordingly, for different i_{s_peak} strengths, e.g., somatic *vs.* axonal, the receptive field of the same electrode is actually different. For example, on the immediate somatic surface of a mouse barrel cortex L5 pyramidal neuron with a 10µm equivalent spherical radius, the peak eAP current i_{s_peak} is estimated to be 1.70 nA; thus, for an electrode recording system (e.g., a square-column microelectrode with a side width of 30µm) with $\delta_{peak} = 5\mu V$, $|Z_{in}| = 20$ MΩ at 1 kHz, $|Z'_e| = 8$ kΩ at 1 kHz, and $\sigma = 2.8$ mS/cm in the rat barrel cortex (Goto et al. 2010), according to Eq. (11.2), $\Delta v_{X\_min} = 10.00\mu V$, and according to Eq. (11.4), $r_{d\_max} = 96.6\mu m$ in the barrel cortex. Thus, a smaller electrode coupled with a noisier recording system would have a shallower receptive field. The receptive fields of a spherical and a square-column electrode are illustrated in Fig. 11.2. The receptive field of a practically implementable square-column electrode approximates to that of an ideal spherical electrode.

Currents along the axon are much smaller (<1%) than the perisomatic current. Currents on the basal dendrites are also much smaller than the perisomatic current (Delgado Ruz and Schultz 2014). With a radius less than 1/10 of the soma, the emission field radius r_{max} of an axonal AP spot is approximately 1/100 of that of the somatic AP. Thus, extracellularly recording an axonal AP is almost impossible unless the axon is attached to the electrode's surface (Patolsky et al. 2006; Bakkum et al. 2013), i.e., within less than 1µm distance. Therefore, MEAs can exclusively be used to map the epsAP current sources in a functional neuronal microcircuit, and AP propagations along individual axons and dendrites can reasonably be ignored (Note however, experimental and simulation results show that axons and dendrites may display minor current source spots when the recording microelectrode is very close to these structures, corresponding to a dipole or multiple source model (Delgado Ruz and Schultz 2014)). In this case, the volumetric current source density is equivalent to the volumetric neuronal density with each monopole equivalent current source corresponding to a neuronal soma.

11.1.4   Temporal Resolution

The temporal resolution of a microelectrode is seldomly talked about, as in most cases, it is assumed to supersede the requirements for recording extracellular neuronal APs. But this assumption needs to be scrutinized for microelectrodes with a side dimension smaller than 10μm, which pick up a much smaller capacitive current through its EDL during neural recording (Guo 2020a). As illustrated in Fig. 11.1, this capacitive current $i_{C_{dl}}(t)$ is passed to charge the parasitic capacitance C_{in} across the amplifier's input terminals after deducting the leakage through the amplifier's input resistance R_{in}, because the actual recording $v_{in}(t)$ is what is built up across the C_{in}. What we are interested is how fast the $v_{in}(t)$ can follow the change of the eAP $v_X(t)$, which is the negative first time derivative of the iAP that has a characteristic frequency of 1 kHz (see Fig. 6.2).

As shown in Fig. 11.1, the overall electrode recording circuit is a second-order system with the following transfer function

$$H(s) = \frac{V_{in}(s)}{V_X(s)} = \frac{\omega_n^2 K s}{s^2 + 2\zeta\omega_n s + \omega_n^2} \qquad (11.5)$$

where the natural frequency $\omega_n = \frac{1}{\sqrt{C_{in}R_{in}C_{dl}(R_s+R_e)}}$, the damping ratio $\zeta = \frac{C_{dl}(R_s+R_e+R_{in})+C_{in}R_{in}}{2\sqrt{C_{in}R_{in}C_{dl}(R_s+R_e)}}$, and the gain $K = C_{dl}R_{in}$ (see Guo (2020b) for derivation). With the necessary neural recording configurations for $C_{dl} \gg C_{in}$ and $R_{in} \gg R_s + R_e$, $\zeta \gg 1$; thus, the system is overdamped and $H(s)$ can be decomposed into a cascade of one first-order highpass filter ($\tau_{hp} \approx 2C_{in}R_{in}$) and one first-order lowpass filter ($\tau_{lp} \approx C_{in}(R_s + R_e)$), with an overall gain close to 1. For example, with $C_{in} = 12$ pF, $R_{in} = 13$ MΩ, $R_s = 33$ kΩ in the neocortex, and $R_e \approx 0$, $\tau_{hp} \approx 312$μs and $\tau_{lp} \approx 0.4$μs.

The capacitive current $i_{C_{dl}}(t)$ sensed by the electrode's C_{dl} in response to variation of $v_X(t)$ minus the leakage through R_{in} is used to charge or discharge C_{in} to build up the recording $v_{in}(t)$. Thus, for fast charging of C_{in} to keep track of the changing $v_X(t)$ (characteristic frequency >2 kHz), $i_{C_{dl}}(t)$ needs to be large with a slow attenuation, i.e., requiring (1) a large C_{dl} to give a large $i_{C_{dl}}(t)$ according to $i_{C_{dl}}(t) = C_{dl}\frac{dv_{C_{dl}}(t)}{dt}$ for a give variation of $v_X(t)$, and (2) a large R_{in} to reduce the leakage, which will require a large τ_{hp} (i.e., a small cutoff frequency as $f_{c\_hp} = \frac{1}{2\pi\tau_{hp}}$) to slow down the attenuation. Meanwhile, C_{in} needs to be small with a τ_{lp} much smaller than 0.25 ms (half width of the eAP spike, see Fig. 6.2). Therefore, C_{dl} and R_{in} need to be large, and C_{in} and R_e need to be small.

The unit step response of this system is

$$s(t) = \frac{K\omega_n}{2\sqrt{\zeta^2 - 1}}(e^{s_2 t} - e^{s_1 t})u(t) \qquad (11.6)$$

where $s_1 = -\omega_n\left(\zeta + \sqrt{\zeta^2 - 1}\right)$ and $s_2 = -\omega_n\left(\zeta - \sqrt{\zeta^2 - 1}\right)$ are the roots to $s^2 + 2\zeta\omega_n s + \omega_n^2 = 0$, i.e., the poles of $H(s)$ in Eq. (11.5). Thus, we can define the *temporal resolution* of this overdamped second-order system to be the peak time t_{peak} required for the unit step response $s(t)$ to reach to the first peak of the overshoot, then

$$t_{peak} = \frac{\ln\frac{s_1}{s_2}}{s_2 - s_1} \approx C_{in}(R_s + R_e)\ln\frac{2R_{in}}{R_s + R_e} = \tau_{lp}\ln\frac{\tau_{hp}}{\tau_{lp}} \qquad (11.7)$$

Using the values for τ_{lp} and τ_{hp} from the above example, we have $t_{peak} = 2.66\mu s$, which is much smaller than 0.25 ms. Thus, the temporal resolution of the electrode recording circuit is independent from the electrode's electrochemical impedance as long as $C_{dl} \gg C_{in}$ and $R_{in} \gg R_s + R_e$.

11.2   An Electrode Unit

As illustrated in Fig. 11.3 inset, a basic electrode unit for mapping the 3D location of an epsAP current source comprises four electrodes (similar to a tetrode in Chap. 10), three in the x–y plane: $e_{<x,\ y,\ z>}000$ with relative coordinates $(0, 0, 0)$, $e_{<x,\ y,\ z>}100$ with $(d_x, 0, 0)$, and $e_{<x,\ y,\ z>}010$ with $(0, d_y, 0)$, and an additional one on the z-axis: $e_{<x,\ y,\ z>}001$ with $(0, 0, d_z)$. The subscript $<x, y, z>$ indicates the indices of the cubic *spatial unit of analysis* (SUA). d_x and d_y are the lateral electrode spacings and commonly equal to each other; and d_z is the vertical electrode spacing and can

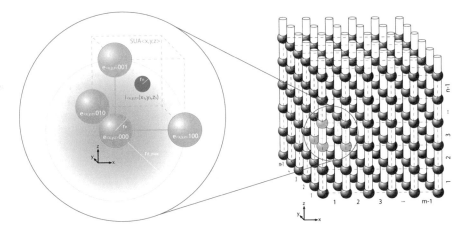

Fig. 11.3   An $m \times m \times n$ ultra-density MEA (right) and a basic electrode unit (left inset). In this example, $d_x = d_y = d_z = d$, and r_{d_max} is set between $\left[d - r_e, \sqrt{3}d - r_e\right]$ as a tradeoff between aliasing and blind zones. Only the receptive field of $e_{<x,\ y,\ z>}000$ is illustrated with $r_{d\_max} = d - r_e$ in the inset. Adapted with permission from (Guo 2020b). Copyright © 2020 IOP Publishing Ltd

also equal to the lateral spacings. The coordinates $i_{<x, \ y, \ z>}(x_s, y_s, z_s)$ of the epsAP current source to be mapped are relative to $e_{<x, \ y, \ z>}000$, where the origin of temporary Cartesian coordinate system is set. The following analysis is based on an equivalent spherical neuronal soma with radius r_n and ideal spherical electrodes with radius r_e, where their 3D coordinates above are set at the spheres' centers. The overlapped receptive fields of each electrode are essential for mapping the 3D coordinates of the epsAP current source. For the current source $i_{<x, \ y, \ z>}(x_s, y_s, z_s)$ in the overlapped receptive fields of these four electrodes, all electrodes can record a potential $v_{in \_ ei}(t)$, with the corresponding deconvoluted eAP $v_{x\_ei}(t) = \frac{i_s(t)}{4\pi\sigma(r_i - r_e)}$ according to Eq. (11.1) (note, deconvolution is not necessary for resolving the source location, see Sect. 10.1.4), where $i = 000, 100, 010,$ or 001 is the electrode's index and r_i is the center-to-center distance between the soma and electrode. The ratios of $e000$'s amplitude to those of others' are $\frac{v_{x\_e000\_peak}}{v_{x\_ei\_peak}} = \frac{r_i - r_e}{r_{000} - r_e} = \alpha_{000/i}$, and $r_i = \alpha_{000/}$ $_i r_{000} + (1 - \alpha_{000/i})r_e$, where $i = 100, 010,$ or 001. Thus, we have Eq. (11.8), which are ultimately reduced to a quartic equation of r_{000} (LeChasseur et al. 2011) that can be solved using the MATLAB function $roots()$, to completely solve the 3D coordinates $i_{<x, \ y, \ z>}(x_s, y_s, z_s)$ of the current source (Guo 2020b), noting that $0 < x_s < d$, $0 < y_s < d, 0 < z_s < d$, and $r_e < r_{000} \leq r_{d \_ max} + r_e$:

$$
\begin{cases}
x_s^2 + y_s^2 + z_s^2 = r_{000}^2 \\
(x_s - d)^2 + y_s^2 + z_s^2 = \left(\alpha_{000/100} r_{000} + \left(1 - \alpha_{000/100}\right)r_e\right)^2 \\
x_s^2 + (y_s - d)^2 + z_s^2 = \left(\alpha_{000/010} r_{000} + \left(1 - \alpha_{000/010}\right)r_e\right)^2 \\
x_s^2 + y_s^2 + (z_s - d)^2 = \left(\alpha_{000/001} r_{000} + \left(1 - \alpha_{000/001}\right)r_e\right)^2
\end{cases}
\tag{11.8}
$$

The resolved 3D coordinates $i_{<x, \ y, \ z>}(x_s, y_s, z_s)$ are the center location of an equivalent spherical neuronal soma with radius r_n (a pyramidal neuron can be mapped to such an equivalent spherical neuron with r_n equal to the center-to-apex distance). Also, note that with the setting of the receptive field $r_{d \_ max} = d - r_e$ in Fig. 11.3 inset, the overlapped receptive field of the four electrodes has a blind zone within the cubic SUA. Nonetheless, by using a proper combination of four electrodes that all bear a recording, any epsAP source fall within the SUA can be resolved.

11.3 Definition of Ultra-Density MEA

Based on the above analyses, the ultra-density MEA as illustrated in Fig. 11.3 is defined with the following three specifications:

1. The electrodes have comparable dimensions to those of the neuronal soma and with proper Z_{in} and δ_{peak} of the recording circuit, so that according to Eqs. (11.2) and (11.4), their receptive fields are small enough to allow (2).

2. The electrode spacings are set so that each SUA contains no more than one active neuronal soma.
3. All active, nonconcurrent epsAP sources in the MEA's recording volume can be resolved spatiotemporally.

In Fig. 11.3, all the electrodes are spherical with a radius of r_e. The ultra-density MEA has $m \times m$ electrodes in the x–y plane, and n electrodes along the z-axis. Along the x, y, and z directions, the electrodes are all evenly spaced with $d_x = d_y = d_z = d$ and centered at the vertices of each cubic SUA. The center-to-center inter-electrode spacing d is set so that $r_{d\_max} \in [d_x - r_e, \sqrt{3}d_x - r_e]$ to make the receptive field of one electrode cover those of the three other adjacent electrodes in each quadrant. The purpose of this arrangement is to make sure that (1) each SUA is covered by one full electrode unit; (2) the redundancies of coverage between adjacent electrode units are minimized; and (3) the 3D coordinates of the center of each electrode are simple.

It is noted that larger cortical neurons have a larger spacing and also result in a larger r_{d_max}, so that the electrode size and spacing are set accordingly to match the neuronal microarchitecture. An advantage of using square-column microelectrodes in the lattice arrangement shown in Fig. 11.3 with $d = r_{d\_max} + \frac{d_e}{2}$ is that the aliasing effect between adjacent SUAs can be mitigated.

11.4 Neural Resolving Power of MEA

The word "bandwidth" is frequently borrowed from the information theory to describe the information throughput of a neural interface in BCIs; however, there is no clear definition for it in such a specific application scenario.

To quantify the resolving power of an MEA, I propose to use four criteria: recording volume V, current source resolution (CSR) κ, percentage p of spatially resolved current sources within V, and spatial resolution Δr_{min} of resolved current source location. The recording volume V is the overall 3D cortical volume covered by the implanted microelectrode grid. The type of recordable extracellular current sources depends on the CSR κ of the recording technology, which includes current sources for, in increment of resolution, LFPs, APs on somas, subthreshold potentials on somas, APs on axons, and subthreshold potentials on axons and dendrites (Fig. 11.4). Current source density (CSD) $d(\kappa, t, x, y, z)$ within V is thus a multivariate function of CSR κ, time t, and space (x, y, z). For a certain type of MEA and its associated recording circuit, κ is fixed; and over a specific time period t, the total number of current sources within V is $N_s(\kappa, t) = \oint_V d(\kappa, t, x, y, z)dV$. The percentage

score p is used to describe the percentage of resolvable current sources at κ within V, which can be *independently isolated* and *spatially resolved* by the MEA. The spatial resolution Δr_{min} of resolved current source location should be the largest/poorest Δr_{min} among all the microelectrodes that involved in resolving the 3D coordinates (x_s, y_s, z_s) of the current source, with $\Delta r_{min} \leq \frac{r_{d\_max}}{2}$ (see Sect. 11.1.2).

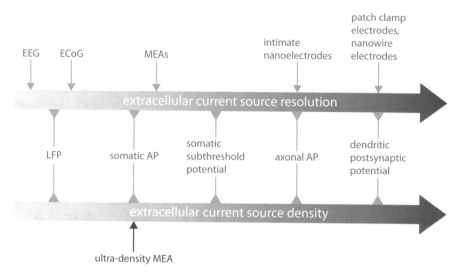

Fig. 11.4 Resolution of extracellular electrical neural recording technologies. Adapted with permission from (Guo 2020b). Copyright © 2020 IOP Publishing Ltd

For conventional extracellular cortical electrodes that do not form a very tight seal to the neuronal membrane, subthreshold potentials cannot be recorded from any part of a neuron (Guo 2020a); neither can the APs on axons (see Sect. 11.1.3); and only the APs on the perisomatic region and/or LFPs can be recorded. Most conventional intracortical microelectrodes can reach to the CSR of perisomatic AP, but only an ultra-density MEA can theoretically detect and resolve all the active perisomatic AP sources within its recording volume V with a $p = 100\%$ (Fig. 11.4). Thus, an ultra-density MEA can have the highest neural information throughput at the somatic AP resolution for intracortical recording that can possibly be implemented by contemporary technologies.

It is noted that for a single intracortical microelectrode, if its recording volume V is considered as its receptive field, it can record all the current sources at its CSR κ within V, however, it can neither independently isolate these sources nor resolve their precise spatial locations (its spatial resolution $\Delta r_{min} = r_{d\_max}$); thus, the recording has a very low neural resolving power.

A further question is even if all the active current sources at a certain CSR within a volume of study are detected and resolved with a $p = 100\%$ from an MEA's recordings, how much neural information is contained in these recordings? It is noted that the higher the CSR κ, the higher the CSD, and the richer neural information is contained in the recordings with a certain p. The ultimate electrical neural information of interest within a certain cortical volume is to fully resolve the spatiotemporal course of electrical neural signal processing in the neural microcircuits, which cannot be fully obtained without reaching to the resolution of dendritic potentials (Fig. 11.4).

In general, the overall *neural information throughput capacity* (NITC) Φ_{max} of a neural interface can be defined as the theoretical maximum number of current sources resolvable by it, which is a multivariate function of the CSR κ of the individual sensors, the recording volume V, the theoretical maximum CSD $d_{max}(\kappa, x, y, z)$, and the percentage $p(\kappa)$ of resovable current sources, with

$$\Phi_{max} = \sum_\kappa \Phi_{max}(\kappa) = \sum_\kappa p(\kappa) \oint_V d_{max}(\kappa, x, y, z)dV \qquad (11.9)$$

where $\Phi_{max}(\kappa)$ is the NITC at a certain CSR κ (e.g., epsAP for an intracortical MEA). The quality of the throughput can be further checked against the spatial resolution Δr_{min} of resolved current source locations.

11.5   Principles of Functional Neural Mapping Using an Ultra-Density MEA

When it comes to an ultra-density MEA, each electrode also records from multiple epsAP sources. Fortunately, solution to this inverse problem is facilitated by rational arrangement of the microelectrodes in the recording volume. The electrode's spacing d is configured to comply with the average spacing of neurons; the electrode's radius r_e or half width $\frac{w_e}{2}$ and its receptive field radius r_{d_max} (primarily affected by the neuron soma size and δ_{peak}) are configured so that $\frac{w_e}{2} + r_{d\_max} = d$. Such a configuration assures to just have at most eight epsAP sources each fall within the eight SUAs surrounding the electrode, respectively, and to resolve all of the individual epsAP sources analytically within its recording volume. We have proposed a recursive approach (Guo 2020b) on how to use an ultra-density MEA to first identify individual epsAP sources each electrode records, and then use Eq. (11.8) to resolve the 3D location of each epsAP source—this is the unique neural resolving capability of ultra-density MEAs comparing to conventional MEAs.

In such an ultra-density MEA, each electrode records from eight SUAs, four upper and four lower; and each SUA is recorded by two complementary independent electrode units located at its vertices. *The key for the ultra-density MEA design is to make sure that each SUA contains no more than one active epsAP source.* Therefore, the temporal and spatial locations of each epsAP source within the recording volume can be precisely mapped.

In the $m \times m \times n$ ultra-density MEA in Fig. 11.3, there are $m \times m \times n$ electrodes and $(m - 1) \times (m - 1) \times (n - 1)$ internal SUAs. Addtionally, there are $m \times m \times 2 + m \times (n - 1) \times 2 + (m - 1) \times (n - 1) \times 2$ extended SUAs surrounding the recording volume. With each internal SUA contains no more than one esAP source, there are maximally $(m - 1) \times (m - 1) \times (n - 1)$ individual epsAP sources within the $m \times m \times n$ electrode grid, i.e., the recording volume V. Each internal

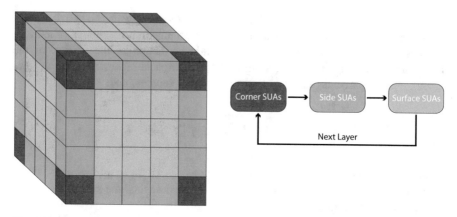

Fig. 11.5 Recursive flowchart for stepwise resolving all internal (i.e., inside the outmost layer) SUAs in the recording volume. Reproduced with permission from (Guo 2020b). Copyright © 2020 IOP Publishing Ltd

epsAP source can be recorded by the eight surrounding electrodes located at the vertices of the cubic SUA.

In Fig. 11.3, if some surrounding epsAP sources exist within an r_{d_max} beyond the recording volume (i.e., in the extended SUAs surrounding the recording volume) of the ultra-density MEA, there is no way to resolve their locations, as they are not recorded by a full electrode unit. However, their recorded spikes can and must be identified using spike sorting and classification. Meanwhile, spikes recorded from the epsAP sources within the peripheral (outmost layer) SUAs (Fig. 11.5) need to first be identified using spike sorting and classification (these two situations are the only cases that spike sorting and classification are required) for these SUAs to resolve their source locations. Afterward, our recursive approach (Guo 2020b) can be applied straightforward to resolve all the rest SUAs.

To completely resolve the epsAP sources within these rest SUAs, the first task is to identify the epsAPs produced by each source in the multiunit recording of each of its eight surrounding electrodes. To achieve this goal, it is assumed that no concurrent firings of adjacent epsAP sources take place, because, otherwise, the individual epsAPs could not be isolated from a superimposed compound AP. Next, Eq. (11.8) can be used to calculate the 3D coordinates of each epsAP source. The flowchart in Fig. 11.5 outlines how all the epsAP sources in their corresponding internal SUAs can be resolved in a recursive way, starting from the four corner SUAs to the side SUAs to the surface SUAs and then to the next inner layer..., until all the internal SUAs are resolved, which is elaborated in details below.

To conquer the first task, in Fig. 11.3, spikes from peripheral SUAs recorded by electrodes at the eight vertices of the cubic internal recording volume (i.e., $e_{<2, 2, 2>}000$, $e_{<m-2, 2, 2>}100$, $e_{<m-2, m-2, 2>}110$, $e_{<2, m-2, 2>}010$, $e_{<2, 2, n-2>}001$, $e_{<m-2, 2, n-2>}101$, $e_{<m-2, m-2, n-2>}111$, and $e_{<2, m-2, n-2>}011$) are already identified by prior spike sorting and classification process, and thus spikes from each corner SUA recorded by each of these electrodes are also identified without using spike sorting. For example, the identified spikes of source $s_{<2, 2, 2>}$ in the recording of $e_{<2, 2, 2>}000$ can be used to identify the same epsAPs in the multiunit recordings of the other seven electrodes on the SUA$<2,2,2>$. Then, the extracted data from one electrode unit include $e_{<2, 2, 2>}000$, $e_{<2, 2, 2>}100$, $e_{<2, 2, 2>}010$ and $e_{<2, 2, 2>}001$ can be used to resolve the 3D coordinates of source $s_{<2, 2, 2>}$ in SUA$<2,2,2>$.

Next, starting from those side SUAs next to the corner SUAs on the 12 sides of the cubic internal recording volume, the esAP sources within these side SUAs can be resolved one by one, as, each time, recordings of the electrodes on each SUA have only one identified epsAP source to resolved. For example, $e_{<3, 2, 2>}000$ (which is also $e_{<2, 2, 2>}100$) records from sources $s_{<2, 2, 2>}$, $s_{<3, 2, 2>}$ and other six sources in the peripheral SUAs resolved earlier assisted with spike sorting. After $s_{<2, 2, 2>}$ is resolved above, the epsAPs from $s_{<3, 2, 2>}$ in the recording of $e_{<3, 2, 2>}000$ can be isolated and then used to resolve $s_{<3, 2, 2>}$ from the electrode unit in SUA$<3,2,2>$.

After all the epsAP sources in the side SUAs are resolved, the epsAP sources in the surface SUAs can be resolved sequentially starting from the SUAs at the corners. For example, now the epsAPs from source $s_{<3, 2, 3>}$ in the recording of $e_{<3, 2, 3>}000$ can be identified, since the epsAPs generated by other sources in the multiunit recording of $e_{<3, 2, 3>}000$ have all been identified previously. Therefore, $s_{<3, 2, 3>}$ can be resolved using the electrode unit origined at $e_{<3, 2, 3>}000$.

After all the surface SUAs are resolved, we can focus on the rest of the recording volume by "peeling off" these surface SUAs. The same steps above are repeated until all the SUAs are resolved.

11.6 Discussions

11.6.1 Particular Issues of Ultra-Density MEA

Generally speaking, the ultra-density MEA technology can be viewed as a careful scaling up of the tetrode approach in terms of size, number, and 3D arrangement of the microelectrodes, with complete and accurate identification and localization of all active epsAP sources within the recording volume in mind. It addresses the limitations of a "pseudo" 3D MEA formed by multiple stepping tetrodes inserted in parallel in a cortical region through an integrated 3D ultra-density MEA. However, it also introduces its own intrinsic problems. In our above analyses, the following four effects have been ignored so far, in order to derive the core principles to this inverse problem. Nonetheless, they are inevitable issues that need to be taken care of properly to develop a sophisticated practical algorithm.

The Aliasing Effect

The extent of aliasing depends on the value of β in $r_{d\_max} + r_e = \beta d$, and counteracts with the blind zone of an electrode unit within an SUA. When $\beta = 1$, there is no aliasing between adjacent SUAs, however, the overlapped receptive field of any electrode unit formed in the SUA has a very limited volume, i.e., a large blind zone; in contrast, when $\beta = \sqrt{3}$, there is no blind zone in any SUA, but the aliasing across adjacent SUAs is the worst.

It is noted that such an aliasing effect is universal to the ultra-density MEA design, no matter how the electrodes are arranged spatially. For example, the triangular or honeycomb arrangement of electrode grid proposed by Kleinfeld, et al. (Kleinfeld et al. 2019) also suffers from this problem.

In design of the ultra-density MEA, β should be chosen in the lower range of $[1, \sqrt{3}]$ to minimize the aliasing effect, while the blind zone effect within certain SUAs can be worked around by choosing a different electrode unit that encompasses the neuron, i.e., all the four electrodes can record from the neuron. For example, using square-column microelectrodes in an ultra-density MEA design with $d = r_{d\_max} + \frac{d_e}{2}$ can mitigate the aliasing effect between adjacent SUAs.

The Peripheral Source Effect

In practical application, if the m and n of the ultra-density MEA are very large for large-scale mapping, chances exist for the extended SUAs surrounding the recording volume to have few to none neurons. When this condition cannot be met, a spike sorting and classification algorithm needs to be used to identify the spikes from epsAP sources in the extended SUAs.

The Interposing Effect

In our above analyses, we did not consider the case when a neuron interposes between two or more adjacent SUAs. In such a case, the number of epsAP sources actually becomes reduced, and the redundancy of recordings across adjacent electrode units can provide sufficient information to resolve this neuron.

Concurrent AP Firing

The neuronal distribution in the cortex is not homogeneous, but hierarchical, and local neural activities are often synchronized. The implantation site should be carefully selected based on prior anatomical knowledge. Thus, the assumption that adjacent SUAs do not have their neurons fire AP concurrently may be violated. Depending on the cortical region and neuronal circuit function, some adjacent neurons may fire concurrently. In such a case, it would be hard to resolve the spatial locations of the individual concurrently firing and spatially adjacent neurons. However, if we treat these neurons as a single epsAP source, its equivalent spatial location can be resolved similarly. For neurons located in nonadjacent SUAs, it does not matter to the inverse problem whether they fire AP concurrently or not. Additionally, LFPs can be recorded by each electrode of the ultra-density MEA, but they carried different information and can be filtered from the multiunit data for separate analysis.

11.6.2 Spatial Oversampling

Because intracortical extracellular microelectrodes can only record epsAPs and the average cortical neuronal spacing is 20μm (Kleinfeld et al. 2019; Mechler et al. 2011), as the electrode pitch d becomes smaller than $r_{d\_max} + r_e$ (see Fig. 11.3 inset), the aliasing effects start to emerge; the further reduction of d, the more severe the aliasing effects become, which consequently poses a higher difficulty in resolving the inverse problem, though the increased density of the electrode unit would help to improve the spatial resolution Δr_{min} of the resolved soma locations, as each electrode becomes closer to the epsAP source. Therefore, practically, there exists a lower boundary below $r_{d\_max} + r_e$ for the electrode pitch d considering the aliasing effects, where the limited gain in spatial resolution Δr_{min} is overwhelmed by the difficulty casued by the aliasing effects in isolation and identification of recorded spikes from each adjacent epsAP sources, not even to consider the difficulties in device fabrication, implantation, and tissue integration for an ultra-density MEA of a higher density.

11.6.3 Implications to BCIs

BCI has a different goal from neuroscience. While neuroscience is interested in mapping the spatiotemporal distributions of electrical signal flows manifested by a cortical neuronal circuit, BCI seeks to identify a spatiotemporal neural activity pattern correlating (may not be the direct control signal) to a certain behavioral or cognitive task to command an external system. Then, some interesting questions are: (1) What is the spatial resolution of the neural functional pattern? (2) Does the BCI electrode array have the sufficient spatial sampling rate to uncover the profile of this functional pattern? (3) If not, what is got there?

Based on our above analyses, it is clear that MEAs, e.g. the UEA (Fig. 1.1a), used in BCIs do not have the sufficient spatial sampling frequency to uncover the full functional pattern; and frequently, they are working at a spatial sampling frequency far below the resolution of the neural functional pattern and under situations where the topology of the underlying neural circuit is not clearly known. Fortunately, a BCI does not need to monitor the entire computation pattern of an underlying neural circuit, as long as a basic pattern that strongly correlates to the planning and/or execution of a behavioral/cognitive task can be extracted from the recordings. However, the higher resolution of the computation pattern gets sampled, the more sophistication of the BCI's performance can be. Thus, there is a dependence of the BCI's sophistication, response speed, and accuracy on the spatial sampling rate of the underlying neural computation pattern. With the significant undersampling of contemporary MEAs used in BCIs, it is no wonder that current BCI research is still at an early stage of development with poor performances on sophistication, response speed, and accuracy.

Then, what spatial sampling frequency should future BCIs aim for? The answer is likely to depend on the target neuronal circuit. A least pattern requiring a minimum number of recording sites is what is needed in a chronic BCI. Thus, the requirement by BCIs on the MEA is much less stringent than that required by neuroscience investigations to fully uncover the spatiotemporal distributions of electrical signal flows manifested by a cortical neuronal circuit, making the chronic implantation practically more achievable.

11.7 Summary

- Functional recording/imaging at the neuronal circuit level aims to identify the association (and hopefully the causal relationship) between a system-level function and the signal flow pattern of neuronal computation within a neuronal microcircuit. Furthermore, to decipher how a neuronal microcircuit functions at the network level with cellular or even subcellular resolution, the neuroscience community calls for clear coupled images on both the network's topology and its spatiotemporally evolving neural activities.
- The signal amplitude resolution of an electrode recording system is determined by Eq. (11.2). Accordingly, a smaller electrode has an unfavorable, lower amplitude resolution (i.e., a larger Δv_{X_min}).
- The spatial resolution of an electrode recording system is determined by Eq. (11.3). Accordingly, a good spatial resolution is achieved at the vicinity of a large current source, e.g. the perisomatic region of a neuron as opposed to its axon or dendrites; and a smaller electrode has a poorer spatial resolution.
- The receptive field of an electrode recording system is determined by Eq. (11.4). Accordingly, for different current source strengths, the receptive field of the same electrode is different; a smaller electrode coupled with a noisier recording system would have a shallower receptive field; and MEAs can only be used to map the somatic AP current sources in a neural microcircuit, and AP propagation along individual axons cannot be detected.
- The overall electrode recording circuit is an overdamped second-order system, the temporal resolution of this system is defined as the peak time t_{peak} required for the unit step response $s(t)$ to reach to the first peak of the overshoot. Generally, the temporal resolution t_{peak} of most electrode recording circuits meets the requirement for recording extracellular neuronal APs. The t_{peak} of the electrode recording circuit is independent from the electrode's electrochemical impedance, as

long as $C_{dl} \gg C_{in}$ and $R_{in} \gg R_s + R_e$. For a smaller t_{peak}, C_{in} and R_e need to be small.

- A basic electrode unit for mapping the 3D location of an epsAP current source comprises four electrodes (similar to a tetrode), three in the x–y plane and an additional one on the z-axis. The overlapped receptive field of each electrode is essential for mapping the 3D coordinates of the epsAP current source. By using a proper combination of four electrodes that all bear a recording, any epsAP source fall within an SUA can be resolved.
- The key for the ultra-density MEA design is to make sure that each SUA contains no more than one active epsAP source. The rational arrangement of an ultra-density MEA is to make sure that: (1) each SUA is covered by one full electrode unit; (2) the redundancies of coverage between adjacent electrode units are minimized; and (3) the 3D coordinates of the center of each electrode are simple.
- To quantify the resolving power of an MEA, four criteria are proposed: recording volume V, CSR κ, percentage p of spatially resolved current sources within V, and spatial resolution Δr_{min} of resolved current source location. An ultra-density MEA can have the highest neural information throughput at the somatic AP resolution for intracortical recording that can possibly be implemented by contemporary technologies.
- A recursive approach is proposed on how to use an ultra-density MEA to first identify individual epsAP sources each electrode records, and then use Eq. (11.8) to resolve the 3D location of each epsAP source—this is the unique neural resolving capability of ultra-density MEAs comparing to conventional MEAs.
- Particular issues of ultra-density MEAs include the aliasing effect, the blind zone effect, the peripheral source effect, the interposing effect, and the concurrent AP firing.
- Practically, there exists a lower boundary below $r_{d\_max} + r_e$ for the electrode pitch d considering the aliasing effects, where the limited gain in spatial resolution Δr_{min} is overwhelmed by the difficulty caused by the aliasing effects in isolation and identification of recorded spikes from each adjacent epsAP source.
- BCI seeks to identify a spatiotemporal neural activity pattern correlating (may not be the direct control signal) to a certain behavioral or cognitive task to command an external system. A BCI does not need to monitor the entire computation pattern of an underlying neural circuit, as long as a basic pattern that strongly correlates to the planning and/or execution of a behavioral/cognitive task can be extracted from the recordings. However, the higher resolution of the computation pattern gets sampled, the more sophistication of the BCI's performance can be. Thus, there is a dependence of the BCI's sophistication, response speed, and accuracy on the spatial sampling rate of the underlying neural computation pattern. A least pattern requiring a minimum number of recording sites is what is needed in a chronic BCI.

Note: This chapter was adapted from (Guo 2020b) with permission.

Exercises

11.1 Answer:

  (a) Why are MEAs needed in neural recording?
  (b) Why are high-density MEAs needed in neural recording?
  (c) What is the ultimate role that an MEA can play in neural recording?

11.2 Answer:

  (a) What factors affect the amplitude resolution of an electrode?
  (b) How is the spatial resolution of an electrode defined? What factors affect it?
  (c) How is the receptive field of an electrode defined? What factors affect it?
  (d) How is the temporal resolution of an electrode defined? Do we need to worry about the temporal resolution of an electrode in general? Why?

11.3 What is an electrode unit? What is it used for? How should an electrode unit be configured?

11.4 What is the definition of an ultra-density MEA? What is the purpose of such an MEA? What is the key for the ultra-density MEA design? What are the particular issues of ultra-density MEAs?

11.5 How to quantify the resolving power of an MEA? What is the resolving power of an ultra-density MEA?

11.6 Describe how an ultra-density MEA should be used in functional neural mapping.

11.7 Regarding to functional recording at the neuronal circuit level,

  (a) What is the goal of neuroscience investigation?
  (b) What is the goal of BCI?
  (c) What are the neural recording requirements for neuroscience and BCI, respectively?

References

Bakkum DJ, Frey U, Radivojevic M, Russell TL, Muller J, Fiscella M, Takahashi H, Hierlemann A (2013) Tracking axonal action potential propagation on a high-density microelectrode array across hundreds of sites. Nat Commun 4. doi: https://doi.org/10.1038/ncomms3181

Blanche TJ, Spacek MA, Hetke JF, Swindale NV (2005) Polytrodes: High-density silicon electrode arrays for large-scale multiunit recording. Journal of neurophysiology 93 (5):2987-3000. doi: https://doi.org/10.1152/jn.01023.2004

Delgado Ruz I, Schultz SR (2014) Localising and classifying neurons from high density MEA recordings. J Neurosci Methods 233:115-128. doi:https://doi.org/10.1016/j.jneumeth.2014.05.037

Du JG, Blanche TJ, Harrison RR, Lester HA, Masmanidis SC (2011) Multiplexed, High Density Electrophysiology with Nanofabricated Neural Probes. PloS one 6 (10). doi: https://doi.org/10.1371/journal.pone.0026204

Goto T, Hatanaka R, Ogawa T, Sumiyoshi A, Riera J, Kawashima R (2010) An Evaluation of the Conductivity Profile in the Somatosensory Barrel Cortex of Wistar Rats. Journal of neurophysiology 104 (6):3388-3412. doi:https://doi.org/10.1152/jn.00122.2010

Guo L (2020a) Perspectives on electrical neural recording: a revisit to the fundamental concepts. J Neural Eng. doi:https://doi.org/10.1088/1741-2552/ab702f

Guo L (2020b) Principles of functional neural mapping using an intracortical ultra-density microelectrode array (ultra-density MEA). J Neural Eng 17. doi:https://doi.org/10.1088/1741-2552/ab8fc5

Jun JJ, Steinmetz NA, Siegle JH, Denman DJ, Bauza M, Barbarits B, Lee AK, Anastassiou CA, Andrei A, Aydin C, Barbic M, Blanche TJ, Bonin V, Couto J, Dutta B, Gratiy SL, Gutnisky DA, Hausser M, Karsh B, Ledochowitsch P, Lopez CM, Mitelut C, Musa S, Okun M, Pachitariu M, Putzeys J, Rich PD, Rossant C, Sun WL, Svoboda K, Carandini M, Harris KD, Koch C, O'Keefe J, Harris TD (2017) Fully integrated silicon probes for high-density recording of neural activity. Nature 551 (7679):232. doi:https://doi.org/10.1038/nature24636

Kleinfeld D, Luan L, Mitra PP, Robinson JT, Sarpeshkar R, Shepard K, Xie C, Harris TD (2019) Can One Concurrently Record Electrical Spikes from Every Neuron in a Mammalian Brain? Neuron 103 (6):1005-1015. doi:https://doi.org/10.1016/j.neuron.2019.08.011

Knopfel T, Diez-Garcia J, Akemann W (2006) Optical probing of neuronal circuit dynamics: genetically encoded versus classical fluorescent sensors. Trends in Neurosciences 29 (3):160-166. doi:https://doi.org/10.1016/j.tins.2006.01.004

LeChasseur Y, Dufour S, Lavertu G, Bories C, Deschenes M, Vallee R, De Koninck Y (2011) A microprobe for parallel optical and electrical recordings from single neurons in vivo. Nat Methods 8 (4):319-325. doi:https://doi.org/10.1038/nmeth.1572

Luan L, Wei XL, Zhao ZT, Siegel JJ, Potnis O, Tuppen CA, Lin SQ, Kazmi S, Fowler RA, Holloway S, Dunn AK, Chitwood RA, Xie C (2017) Ultraflexible nanoelectronic probes form reliable, glial scar-free neural integration. Science Advances 3 (2). doi: https://doi.org/10.1126/sciadv.1601966

Mechler F, Victor JD, Ohiorhenuan I, Schmid AM, Hu Q (2011) Three-dimensional localization of neurons in cortical tetrode recordings. Journal of neurophysiology 106 (2):828-848. doi:https://doi.org/10.1152/jn.00515.2010

Patolsky F, Timko BP, Yu GH, Fang Y, Greytak AB, Zheng GF, Lieber CM (2006) Detection, stimulation, and inhibition of neuronal signals with high-density nanowire transistor arrays. Science 313 (5790):1100-1104. doi:https://doi.org/10.1126/science.1128640

Peterka DS, Takahashi H, Yuste R (2011) Imaging Voltage in Neurons. Neuron 69 (1):9-21. doi: https://doi.org/10.1016/j.neuron.2010.12.010

Rios G, Lubenov EV, Chi D, Roukes ML, Siapas AG (2016) Nanofabricated Neural Probes for Dense 3-D Recordings of Brain Activity. Nano Lett 16 (11):6857-6862. doi:https://doi.org/10.1021/acs.nanolett.6b02673

Rivnay J, Wang HL, Fenno L, Deisseroth K, Malliaras GG (2017) Next-generation probes, particles, and proteins for neural interfacing, Science Advances 3 (6). doi:https://doi.org/10.1126/sciadv.1601649

Scholvin J, Kinney JP, Bernstein JG, Moore-Kochlacs C, Kopell N, Fonstad CG, Boyden ES (2016) Close-Packed Silicon Microelectrodes for Scalable Spatially Oversampled Neural Recording. IEEE T Bio-Med Eng 63 (1):120-130. doi:https://doi.org/10.1109/Tbme.2015.2406113

Wang Y, Zhu H, Yang H, Argall AD, Luan L, Xie C, Guo L (2018) Nano functional neural interfaces. Nano Res. doi:https://doi.org/10.1007/s12274-018-2127-4

Wei XL, Luan L, Zhao ZT, Li X, Zhu HL, Potnis O, Xie C (2018) Nanofabricated Ultraflexible Electrode Arrays for High-Density Intracortical Recording. Advanced Science 5 (6). doi:https://doi.org/10.1002/advs.201700625

Part III
Principles of Electrical Neural Stimulation

Chapter 12
Neuronal Stimulation

Electrical neural stimulation is a complementary technique to electrical neural recording. Both techniques use electrodes to interface with the neuron or neural tissue. However, as the current flows in the opposite direction in the electrode circuits of these two modalities and the stimulating current is orders of magnitude larger than the recording current, requirements for both the electrode designs and associated electronic circuits are quite different. Chapter 4 has discussed characteristics of the stimulating electrode and the circuit. Furthermore, during electrical neural recording, the neuron is viewed as a current source; in contrast, during electrical neural stimulation, the neuron is viewed as a passive load during the initial subthreshold depolarization. Following similar analyses as in Chaps. 5 and 6, this chapter uses equivalent electrical circuit modeling and frequency-domain analyses to reveal the neuronal stimulation mechanisms under different configurations, including intracellular stimulation, extracellular open-field stimulation, and extracellular substrate stimulation with and without electroporation.

12.1 Intracellular Stimulation

Intracellular neural stimulation, as performed using a glass micropipette or whole-cell patch clamp electrode (Fig. 12.1) to inject a positive current $I_{stim}(s)$ into the cytoplasma to depolarize the membrane, can be modeled as using $I_{stim}(s)$ to charge the transmebrane capacitance C_m through the parallel RC circuit, whose volatge $V_m(s)$ (time-domain representation is $v_m(t)$) is the AC transmembrane voltage. Considering the leakage through the seal R_{seal} between the membrane and glass micropipette, $V_m(s)$ is attenuated to $V'_m(s)$. When the peak amplitude of $v'_m(t)$ reaches to the AP threshold, an AP will be initiated in the neuron. With the GND placed immediately outside the neuron, we have

© The Author(s), under exclusive license to Springer Nature Switzerland AG 2022
L. Guo, *Principles of Electrical Neural Interfacing*,
https://doi.org/10.1007/978-3-030-77677-0_12

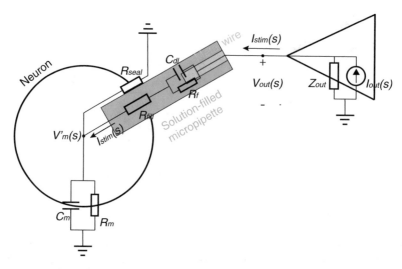

Fig. 12.1 Intracellular neural stimulation using a glass micropipette or whole-cell patch clamp electrode

$$V'_m(s) = \frac{R_m + R_{seal}}{sC_m(R_m + R_{seal}) + 1} I_{stim}(s) \qquad (12.1)$$

The transfer function $H(s) = \frac{R_m + R_{seal}}{sC_m(R_m + R_{seal}) + 1}$ is a first-order lowpass filter with a -3 dB cutoff frequency of $f_c = \frac{1}{2\pi C_m(R_m + R_{seal})}$ and a passband gain of $R_m + R_{seal}$. Interpretation of this overall stimulating circuit is straightforward, and the time-domain analyses can be derived accordingly (Plonsey and Barr 2007). Please see Exercise 12.1.

12.2   Extracellular Stimulation: Basic Relationships Between eFP and Transmembrane Voltage

12.2.1   Intimate Stimulation

The situation in extracellular neural stimulation is more complicated. Let us first imagine a spherical neuron of radius r_n suspended in an infinite homogeneous electrolyte, whose plasma membrane is passive during subthreshold activities and is modeled as a parallel resistor R_m and a capacitor C_m, as shown in Fig. 12.2a. A spherical electrode of radius r_e is placed in the electrolyte so that its surface transverses at a Point X immediately outside the neuron. The electrode delivers an AC current $I_{stim}(s)$. Let us first assume $r_e \gg r_n$, so that the neuronal membrane can be uniformly charged by the spreading ionic current. This ionic $I_{stim}(s)$ has two parallel paths to dissipate: (1) through the spreading resistance R_s of the electrolyte to the

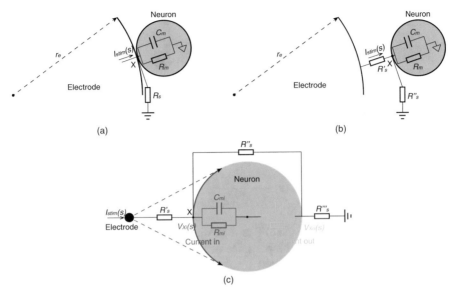

Fig. 12.2 Extracellular neural stimulation in an open medium. (**a**) and (**b**) The electrode's radius is much larger than that of the neuron, i.e., $r_e \gg r_n$, and it is assumed that the neuronal membrane is uniformly charged by the spreading ionic current. (**a**) The neuron is placed in intimate contact with the electrode at Point X. (**b**) The neuron is located at a Point X with a distance r_0 ($r_0 > r_e$) away from the center of the electrode. (**c**) The electrode radius is much smaller than that of the neuron, i.e., $r_e \ll r_n$, and the neuronal membrane is nonuniformly charged by the spreading ionic current, with the red membrane *mi* corresponding to where the current goes into the cell and the green membrane *mo* corresponding to where the same current gets out. The spreading resistance is splitted into three portions with $R_s = R_s' + R_s'' + R_s'''$. The average eFP generated by $I_{stim}(s)$ through the electrode on the *mi* membrane surface is denoted as $V_{Xi}(s)$ and that on the *mo* membrane surface as $V_{Xo}(s)$. $Z_{mi} = C_{mi} \| R_{mi}$, $Z_{mo} = C_{mo} \| R_{mo}$, and $Z_m = C_m \| R_m = Z_{mi} \| Z_{mo}$

GND at the infinity, and (2) across the cell membrane to reach to the virtual *GND* located at the center of the neuron. The existence of such a virtual *GND* is a characteristic of the neuronal model (also see Chap. 2, Sect. 2.4). The extracellular voltage at Point X thus is

$$V_X'(s) = I_{stim}(s) \frac{\frac{R_m R_s}{R_m + R_s}}{s C_m \frac{R_m R_s}{R_m + R_s} + 1} = V_{stim}(s) \frac{\frac{R_m}{R_m + R_s}}{s C_m \frac{R_m R_s}{R_m + R_s} + 1} \tag{12.2}$$

where $V_X'(s)$ is the eFP or the *in-situ* stimulating voltage at Point X in presence of the neuron, and $V_{stim}(s) = I_{stim}(s) R_s$ is defined as the *stimulating voltage* in absence of the neuronal load. Note that the transmembrane voltage is $V_m(s) = -V_X'(s)$ according to the configuration and definitions.

The transmembrane current flowing from the extracellular space into the cytoplasma is

$$I_m(s) = I_{stim}(s)\frac{R_s}{\frac{R_m}{sC_mR_m+1} + R_s} \tag{12.3}$$

Then, we have

$$V_m(s) = -I_m(s)\frac{R_m}{sC_mR_m + 1} = -I_{stim}(s)\frac{\frac{R_sR_m}{R_m+R_s}}{1 + sC_m\frac{R_mR_s}{R_m+R_s}}$$

$$= -V_{stim}(s)\frac{\frac{R_m}{R_m+R_s}}{sC_m\frac{R_mR_s}{R_m+R_s} + 1} = -V_X'(s) \tag{12.4}$$

In absence of the neuron, the eFP at the Point X is $V_X(s) = I_{stim}(s)R_s = V_{stim}(s)$, and according to Eq. (12.2), the presence of the neuron in the electrolyte slightly attenuates the eFP $V_X(s)$ generated by the injected current $I_{stim}(s)$ to $V_X'(s)$ by absorbing a tiny amount of $I_{stim}(s)$ into it. Nonetheless, given empirical values of the parameters (see below), $V_X'(s) \approx V_X(s)$ in Eq. (12.2).

Equation (12.4) *is the governing equation for extracellular subthreshold neural stimulation.* The "-" means a negative current or voltage stimulus generates a positive voltage to the inner side of C_m, i.e., causing depolarization to the membrane. The modulating factor $\frac{\frac{R_m}{R_m+R_s}}{sC_m\frac{R_mR_s}{R_m+R_s}+1}$ is a first-order lowpass filter with a -3 dB cutoff frequency of $f_c = \frac{1}{2\pi C_m\frac{R_mR_s}{R_m+R_s}}$ and a passband gain of $G = \frac{R_m}{R_m+R_s}$. With empirical values for $R_s \approx 143\ \Omega$ (e.g., $r_e = 50\mu m$), $R_m = 300\ M\Omega$ and $C_m = 5.3$ pF (e.g., $r_n = 7.5\mu m$), $f_c = 210$ MHz and $G \approx 1$. As the frequencies of conventional neural stimuli (e.g., a $100\mu s$ pulse train) fall within the passband of this lowpass filter, Eq. (12.4) reduces to

$$V_m(s) \approx -V_{stim}(s) = -I_{stim}(s)R_s = -V_X(s) \tag{12.5}$$

12.2.2 Distant Stimulation

When the neuron is located at a Point X with a distance r_0 ($r_0 > r_e$ and $r_e \gg r_n$) away from the center of the stimulating electrode (Fig. 12.2b), following a similar derivation, we have

$$V_m(s) = -I_{stim}(s)\frac{R_s''}{sR_s''C_m + 1} = -V_{stim}(s)\frac{\frac{R_s''}{R_s}}{sR_s''C_m + 1} \tag{12.6}$$

where $R'_s + R''_s = R_s$. The cutoff frequency of this first-order lowpass filter becomes even larger than that in the intimate stimulation configuration in Fig. 12.2a. Thus, the frequencies of conventional neural stimuli also fall within the passband of this lowpass filter, and Eq. (12.6) reduces to

$$V_m(s) = -\frac{R''_s}{R_s} V_{stim}(s) = -\frac{r_e}{r_0} V_{stim}(s) \qquad (12.7)$$

And we have

$$V'_X(s) = I_{stim}(s) \frac{R''_s \frac{1}{sC_m}}{R''_s + \frac{1}{sC_m}} = V_X(s) \frac{1}{sR_sC_m + 1} \qquad (12.8)$$

where $V_X(s) = I_{stim}(s)R''_s$. Thus, we also have $V'_X(s) \approx V_X(s)$ and $V_m(s) = -V'_X(s)$.

Thus, with uniform membrane depolarization, no matter intimate or distant stimulation, the intracellular depolarization $V_m(s)$ is the polarity-inversed eFP $V'_X(s)$ developed around the neuron. In reality, this situation applies to how intracortical LFPs affect neurons immersed in them.

12.2.3 Electrode Much Smaller Than the Neuron

Last, we consider another extreme case in which $r_e \ll r_n$, i.e., the inverse case as those in Sects. 12.2.1 and 12.2.2. When the electrode size is comparable to that of the neuron and nonuniform depolarization needs to be considered, the following analysis still applies. The electrode is placed at a distance r_0 ($r_0 > r_e$) from the Point X on the neuron surface, and the *GND* is placed at the infinity, as illustrated in Fig. 12.2c. The neuronal membrane is nonuniformly depolarized, and the portion of the membrane *mi* (stands for the *membrane* where the current flows *in*) depends on the distance between the electrode and neuron. We have the average *mi* and *mo* (stands for the *membrane* where the current flows *out*) potentials:

$$V_{mi}(s) = -\frac{Z_{mi}}{Z_{mi} + Z_{mo}} (V_{Xi}(s) - V_{Xo}(s)) = -\frac{Z_m}{Z_{mo}} (V_{Xi}(s) - V_{Xo}(s)) \qquad (12.9)$$

$$V_{mo}(s) = \frac{Z_{mo}}{Z_{mi} + Z_{mo}} (V_{Xi}(s) - V_{Xo}(s)) = \frac{Z_m}{Z_{mi}} (V_{Xi}(s) - V_{Xo}(s)) \qquad (12.10)$$

$V_{mi}(s)$ and $V_{mo}(s)$ have opposite polarities. Note that $\left|\frac{Z_m}{Z_{mo}}\right| < 1$, $\left|\frac{Z_m}{Z_{mi}}\right| < 1$ and $|Z_{mi}| > |Z_{mo}|$, thus $|V_{mi}| > |V_{mo}|$ and the neuron is easier to initiate an AP in the *mi* membrane under a negative $I_{stim}(s)$. However, as the $I_{stim}(s)$ flows in the parallel circuit of $Z_{mi} + Z_{mo}$ and R''_s with $R''_s \ll |Z_{mi} + Z_{mo}|$, there is minimum amount of

$I_{stim}(s)$ entering the neuron through mi, and thus the stimulation is very inefficient. Furthermore, it is noted that without a tight seal between the electrode and neuron, the eFP difference $V_{Xi}(s) - V_{Xo}(s)$ generated by the stimulating current $I_{stim}(s)$ through the parallel circuit of $Z_{mi} + Z_{mo}$ and R_s'' is very small, so that the amplidute of $V_{mi}(s)$ is actually much smaller. Fortunately, such inefficiency can be mitigated by creating a tight seal between the electrode and neuron, as will be shown in Sect. 12.3.4.

12.3 Extracellular Stimulation Using a Planar Substrate Microelectrode

This situation applies to electrical stimulation to in vitro cultured cells, as is being prevalently practiced.

12.3.1 Electrode Area Equal to the Neuronal Junctional Area

In the first case, we consider that the electrode surface exactly matches to the neuronal junctional area, as illustrated in Fig. 12.3a. This configuration can be physically achieved by patterning the electrode surface with cell adhesive extracellular matrix (ECM) proteins such as laminin or fibronectin, whereas coating the insulative surface with cell repulsive molecules such as polyethylene terephthalate (Chen et al. 1997). The neuron membrane is thus nonuniformly affected by the underlying stimulating electrode. For convenience, we place the *GND* or counter electrode symmetrically in the vicinity of the neuron such that $R_s = 0$. By solving the equivalent circuit in Fig. 12.3a, we have

$$V_{jm}(s) = -\frac{Z_{jm}}{Z_{jm} + Z_{njm}} V_X'(s) = -\frac{Z_{jm}}{R_{seal} + Z_{jm} + Z_{njm}} R_{seal} I_{stim}(s)$$

$$= -\frac{Z_{jm}}{R_{seal} + Z_{jm} + Z_{njm}} V_{stim}(s) \approx -\frac{\frac{C_{njm}}{C_m}}{s \frac{R_{seal} C_{jm} C_{njm}}{C_m} + 1} V_{stim}(s) \qquad (12.11)$$

and

$$V_{njm}(s) = \frac{Z_{njm}}{Z_{jm} + Z_{njm}} V_X'(s) \qquad (12.12)$$

where $Z_{jm} = \frac{1}{sC_{jm}} \| R_{jm}$, $Z_{njm} = \frac{1}{sC_{njm}} \| R_{njm}$, and $V_{stim}(s) = R_{seal} I_{stim}(s) = V_X(s)$ is the stimulating voltage in absence of the capacitive cell but preserving the junctional seal. The approximation in Eq. (12.11) is reached when the passive membrane

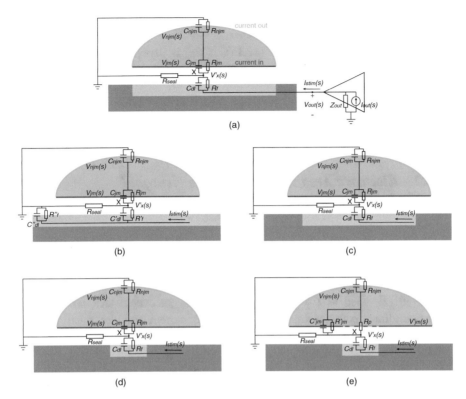

Fig. 12.3 Extracellular neural stimulation of a cultured cell using a planar substrate microelectrode. (**a**) The electrode area equals the neuronal junctional area. (**b**) The electrode area is larger than the neuronal junctional area. (**c**) The electrode area is smaller than the neuronal junctional area. (**d**) The electrode area is much smaller than the neuronal junctional area. (**e**) Extracellular stimulation using a planar substrate microelectrode after electroporation with the electrode area smaller than the neuronal junctional area

resistances R_{jm} and R_{njm} are ignored (Guo 2020). The modulating factor in the approximation in Eq. (12.11) is also a first-order lowpass filter with an empirical -3 dB cutoff frequency $f_c = 1.35$ MHz. Thus, conventional stimulus frequencies also fall within its passband, and Eq. (12.11) reduces to

$$V_{jm}(s) \approx -\frac{C_{njm}}{C_m} V_{stim}(s) \tag{12.13}$$

Similarly,

$$V_{njm}(s) \approx \frac{C_{jm}}{C_m} V_{stim}(s) \tag{12.14}$$

Because $V'_X(s) \approx V_X(s) = V_{stim}(s)$, the majority of $I_{stim}(s)$ leaks through R_{seal}, rather than going into the overlying neuron.

According to Eqs. (12.13) and (12.14), if C_{jm} is depolarized with a negative $I_{stim}(s)$, C_{njm} is hyperpolarized. Thus, the overall cell membrane is nonuniformly polarized. As $V'_X(s)$ can be either negative or positive, either C_{jm} or C_{njm} can be the initiator for an overall AP. But, as $C_{njm} > C_{jm}$, it is more power-efficient to use a negative $I_{stim}(s)$ to induce C_{jm} for an AP.

12.3.2 Electrode Area Larger Than the Neuronal Junctional Area

In the second case, we consider a large planar microelectrode with the overlying neuron only covering a portion of its surface, as illustrated in Fig. 12.3b. This configuration is more commonly seen in neuronal culture stimulations. The geometric electrode surface is divided into the covered (modeled as $C'_{dl} \parallel R'_{dl}$) and the uncovered (modeled as $C''_{dl} \parallel R''_{dl}$) areas, with the uncovered area providing an unwanted direct path to drain the stimulating current. As the overlying cell minimumly affects the current density on the covered electrode surface, the actual current to depolarize the overlying neuron reduces to $\frac{A_{covered}}{A_e} I_{stim}(s)$ (see Exercise 12.3). The configuration then becomes the same as the case in Sect. 12.3.1 but with the reduced stimulating current.

12.3.3 Electrode Area Smaller Than the Neuronal Junctional Area

In the third case, we consider that the electrode area is smaller than the neuronal junctional area such that the electrode is shadowed underneath the neuron, as illustrated in Fig. 12.3c. As $I_{stim}(s)$ is released from the electrode surface, a portion goes into the overlying neuron, the rest leaks through the junctional seal. While in the junctional seal other than over the electrode surface, the lateral current also can go into the neuron to depolarize the overlying membrane. As a result, the overall effect is that the entire junctional membrane is depolarized by the current $I_{stim}(s)$. The situation then is the same as the case in Sect. 12.3.1.

12.3.4 Electrode Area Much Smaller Than the Neuronal Junctional Area

Next, we consider an extreme case in which the electrode area is much smaller than the junctional area, as illustrated in Fig. 12.3d. The analysis is similar as the case in Sect. 12.3.3, and we have the average $V_{jm}(s) \approx -\frac{C_{njm}}{C_m} V_X'(s)$. However, the current density in the junctional area is nonuniform, with a higher current density centered around the electrode while concentrically fading towards the edges of the seal (Buitenweg et al. 2003). At the higher current density area around the electrode, the local membrane is more depolarized/hyperpolarized, so that this piece of membrane is more likely to initiate the AP when sufficient depolarization is caused. Such a much smaller electrode can provide a higher efficiency to induce the AP in the tiny local membrane (Buitenweg et al. 2003), which can then spread over the entire membrane. However, a higher CIC of the stimulating electrode material is required, otherwise, more severe side-effects could happen due to the higher current density with such a small electrode, such as generation of extra reactive ion species and more deterioration to the electrode (see Chap. 4). Nonetheless, C_{jm} still corresponds to the entire junctional area, so that the average $V_{jm}(s) \approx -\frac{C_{njm}}{C_m} V_X'(s)$.

The average $V_{njm}(s) \approx \frac{C_{jm}}{C_m} V_{stim}(s)$ is still silimar to that in Section 12.3.1, and a positive stimulating current $I_{stim}(s)$ can depolarize it to initiate an AP, though the efficiency now is much lower than using a negative stimulating current to depolarize the $V_{jm}(s)$ for an AP.

12.3.5 Extracellular Stimulation Using a Planar Substrate Microelectrode After Electroporation

Last, we consider the configuration in which the electrode is smaller than the overlying cell membrane (Park et al. 2019), as illustrated in Fig. 12.3e. The poration voltage shock delivered through the electrode can actually affect a larger junctional membrane area around the electrode, due to the spreading effect of the voltage through the R_{seal}. As a result, the porated membrane area is larger than the electrode area. When a positive $I_{stim}(s)$ is then delivered through the electrode, a substantial portion of it enters the cell through the porated membrane (modeled as an access resistance R_p), depolarizing C_{njm}. In parallel, similar to the case in Sect. 12.3.4, the portion of $I_{stim}(s)$ leaking through R_{seal} can also partially enter the cell through the intact junctional membrane to hyperpolarize C_{jm} and depolarize C_{njm}. However, as this portion of $I_{stim}(s)$ is much smaller than that enters the cell through R_p, its effect on C_{jm} and C_{njm} is very small. When the $I_{stim}(s)$ is negative, the reverse holds. Thus, either a positive (through C_{njm}) or negative (through C_{jm}) $I_{stim}(s)$ can induce an AP in the cell, but a positive $I_{stim}(s)$ is much more efficient. When a negative $I_{stim}(s)$ is used, the efficiency is comparable to the case in Sect. 12.3.4.

Solving the equivalent circuit in Fig. 12.3e, we have

$$V_{njm}(s) = \frac{Z_{njm}}{Z''_{jm} + Z_{njm}} V'_X(s) \tag{12.15}$$

$$V'_{jm}(s) = -\frac{Z''_{jm}}{Z''_{jm} + Z_{njm}} V'_X(s) \tag{12.16}$$

$$V'_X(s) = \left(\left(Z''_{jm} + Z_{njm} \right) \parallel R_{seal} \right) I_{stim}(s) \tag{12.17}$$

where $Z''_{jm} = Z'_{jm} \parallel R_p = \frac{1}{sC'_{jm}} \parallel R'_{jm} \parallel R_p$.

The magnitude of $V'_{jm}(s)$ is substantially smaller than that of $V_{njm}(s)$ due to the parallel R_p. Thus, *the overall effect of membrane electroporation is to substantially lower the equivalent resistance of the junctional membrane* (as $\left(R_p \parallel R'_{jm} \right) < R_{jm}$). Either a positive or a negative stimulus can induce APs in the neuron, but with distinct efficiencies. When using a negative current stimulus, $V'_{jm}(s)$ is depolarized and the porated stimulation has an even lower power efficiency than the nonporated stimulation in the case in Sect. 12.3.4. In contrast, with a positive current stimulus, $V_{njm}(s)$ is depolarized and the porated stimulation is more efficient than the nonporated stimulation.

However, the majority of $I_{stim}(s)$ still leaks through R_{seal}. This configuration is similar to the intracellular stimulation using a glass or whole-cell patch clamp micropipette in Fig. 12.1. As the R_{seal} here is much smaller than that in Fig. 12.1 and the access resistance R_p is much larger than R_{fill}, such an extracellular stimulation is much less power-efficient.

12.4 Optimizing Stimulation Efficacy

Based on the above analyses on both intracellular and extracellular stimulation configurations, to enable a good efficiency, the following principles can be summarized:

1. The access impedance for the stimulating current to enter the cell needs to be minimized. This can be done by either mechanically break the membrane by a sharp glass micropipette as in Sect. 12.1 or porate the membrane using an underlying electrode as in Sect. 12.3.5.
2. The seal resistance R_{seal} through which the stimulating current leaks needs to be maximized. This can be done by improving the seal between the membrane and

the glass micropipette using a whole-cell patch clamp configuration in intracellular stimulation or increasing the seal between the membrane and electrode substrate by enhancing the cell adhesion using micro/nanostructures (Park et al. 2019; Dipalo et al. 2018; Robinson et al. 2012; Hai et al. 2010; Didier et al. 2020).

3. It is essential to avoid unsealed electrode surface directly exposing to the open medium, as such an exposed surface serves as a direct path to drain the stimulating current.

4. The local current density also matters, as the local peak membrane depolarization is proportional to the local peak current density, and a localized AP onset is sufficient to trigger an AP over the entire cell membrane. To minimize the overall stimulating current, using a smaller electrode with higher current density within the safety limit is preferred.

12.5   Summary

- Intracellular neural stimulation using a glass micropipette or whole-cell patch clamp electrode injects a current into the cytoplasma to charge/discharge the transmembrane capacitance C_m through the parallel RC circuit.
- With uniform membrane depolarization, no matter intimate or distant stimulation, the intracellular depolarization $V_m(s)$ is the polarity-inversed eFP $V'_X(s)$ developed around the neuron. In reality, this situation applies to how intracortical LFPs affect neurons immersed in them.
- When the electrode is much smaller than the neuron, the neuron is easier to initiate an AP in the mi membrane under a negative $I_{stim}(s)$. However, without a tight seal between the electrode and neuron, the induced depolarization is very inefficient.
- When stimulating using a planar substrate microelectrode, according to Eqs. (12.13) and (12.14), if C_{jm} is depolarized with a negative $I_{stim}(s)$, C_{njm} is hyperpolarized. Thus, the overall cell membrane is nonuniformly polarized. As $V'_X(s)$ can be either negative or positive, either C_{jm} or C_{njm} can be the initiator for an overall AP. But, as $C_{njm} > C_{jm}$, it is more power-efficient to use a negative $I_{stim}(s)$ to induce C_{jm} for an AP.
- When the electrode is larger than the neuronal junctional area, the uncovered area provides an unwanted direct path to drain the stimulating current and the effective stimulating current is reduced to $\frac{A_{covered}}{A_e} I_{stim}(s)$.
- When the electrode is much smaller than the neuronal junctional area, the current density in the junctional area is nonuniform, with a higher current density centered around the electrode, where the local membrane is more depolarized. Therefore, a much smaller electrode can provide a higher efficiency to induce the AP in the tiny local membrane, which can then spread over the entire membrane.
- When the overlying membrane is electroporated, a positive $I_{stim}(s)$ is much more efficient to induce an AP in the cell through C_{njm}.

- Based on the analyses, four insightful principles for neural stimulation are summarized.

Exercises

12.1 Referring to Fig. 12.1, an impaled glass micropipette is used to stimulate a neuron intracellularly. The spherical neuron has a diameter of 15μm. $C_m = 5.3$ pF, $R_m = 200$ MΩ. A membrane depolarization of 20 mV is needed for the neuron to fire an AP. $i_{stim}(t)$ is a positive 100μs wide square pulse.

   (a) For $R_{seal} = 20$ MΩ, calculate the minimum amplitude of $i_{stim}(t)$ to initiate an AP in the neuron.
   (b) For $R_{seal} = 2$ GΩ, calculate the minimum amplitude of $i_{stim}(t)$ to initiate an AP in the neuron.
   (c) What conclusion can you reach with regard to how R_{seal} affects the stimulation efficiency.

12.2 For extracellular stimulation in an open medium,

   (a) What is the direct effector of neural stimulation? That is, which quantity directly affects the $v_m(t)$?
   (b) Does the polarity of the stimulus matter?
   (c) Why is stimulation in an open medium not efficient? How to increase the efficiency?

12.3 Referring to Fig. 12.3b, prove that the overlying cell minimally affects the current density on the covered electrode surface, so that the actual current to depolarize the overlying neuron can be represented as $\frac{A_{covered}}{A_e} I_{stim}(s)$.

12.4 For extracellular stimulation using a planar substrate microelectrode,

   (a) How should the electrode size be set?
   (b) What is the problem when the electrode is much larger than the overlying neuron?
   (c) Does the polarity of the stimulus matter? Which stimulus polarity is more efficient and why?
   (d) What is the effect when the membrane is electroporated? Which stimulus polarity is more efficient in this case and why?

12.5 In this problem, we investigate whether a cardiomyocyte can pace a neuron to fire APs. In the figure (a) below, the neuron sits on top of a cardiomyocyte, who fires APs rhythmically at 1 Hz. And in figure (b), the cardiomyocyte wraps on top of the neuron. The spreading resistance $R_s = 1$ kΩ, the seal resistance between the neuron and cardiomyocyte in (a) $R_{sealnc} = 100$ kΩ, the seal resistance between the cardiomyocyte and neuron in (b) $R_{sealcn} = 10$ MΩ, and the seal resistance between the cardiomyocyte and the substrate in (b) $R_{seal} = 1$ MΩ. The specific membrane capacitance is 0.01 pF/μm^2 for both the neuron and cardiomyocyte. The maximum rising slope of a cardiomyocyte AP is $\left[\frac{\Delta v_{mAP}}{\Delta t}\right]_{max} = \frac{110 \text{ mV}}{3 \text{ ms}} = 36.7$ V/s. The diameter of the cardiomyocyte is 500μm. The diameter of the neuron is 10μm. The depolarization threshold of the neuron is 20 mV. In each case, please calculate whether an AP of the cardiomyocyte can trigger an AP in the neuron.

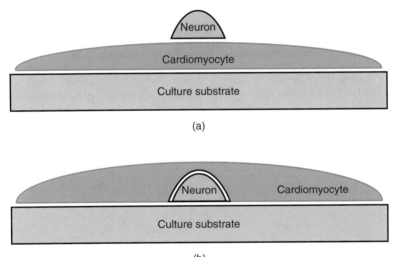

(a)

(b)

12.6 Using a planar substrate microelectrode for long-term extracellular stimulation where the electrode integrity and biosafety need to be considered,

   (a) How should the stimulus waveform be designed? Please consider a two-phase current stimulus. The neuron needs to be activated only in the first phase.
   (b) What are the requirements on the electrode parameters in order to deliver the current stimulus?

References

Buitenweg JR, Rutten WL, Marani E (2003) Geometry-based finite-element modeling of the electrical contact between a cultured neuron and a microelectrode. IEEE T Bio-Med Eng 50 (4):501-509

Chen CS, Mrksich M, Huang S, Whitesides GM, Ingber DE (1997) Geometric control of cell life and death. Science 276 (5317):1425-1428

Didier CM, Kundu A, DeRoo D, Rajaraman S (2020) Development of In Vitro 2D and 3D Microelectrode Arrays and Their Role in Advancing Biomedical Research. Journal of Micromechanics and Microengineering

Dipalo M, McGuire AF, Lou HY, Caprettini V, Melle G, Bruno G, Lubrano C, Matino L, Li X, De Angelis F, Cui B, Santoro F (2018) Cells Adhering to 3D Vertical Nanostructures: Cell Membrane Reshaping without Stable Internalization. Nano Lett 18 (9):6100-6105. doi:https://doi.org/10.1021/acs.nanolett.8b03163

Guo L (2020) Perspectives on electrical neural recording: a revisit to the fundamental concepts. J Neural Eng 17 (1):013001. doi:10.1088/1741-2552/ab702f

Hai A, Shappir J, Spira ME (2010) Long-Term, Multisite, Parallel, In-Cell Recording and Stimulation by an Array of Extracellular Microelectrodes. Journal of neurophysiology 104 (1):559-568. doi:https://doi.org/10.1152/jn.00265.2010

Park JS, Grijalva SI, Jung D, Li S, Junek GV, Chi T, Cho HC, Wang H (2019) Intracellular cardiomyocytes potential recording by planar electrode array and fibroblasts co-culturing on multi-modal CMOS chip. Biosensors and Bioelectronics 144:111626

Plonsey R, Barr RC (2007) Bioelectricity: a quantitative approach. Springer Science & Business Media,

Robinson JT, Jorgolli M, Shalek AK, Yoon MH, Gertner RS, Park H (2012) Vertical nanowire electrode arrays as a scalable platform for intracellular interfacing to neuronal circuits. Nature nanotechnology 7 (3):180-184. doi:https://doi.org/10.1038/nnano.2011.249

Chapter 13
Electrical Stimulation to Promote Neuronal Growth

Electric fields have been observed to promote and guide the neurite outgrowth of neurons. These observations root in the embryonic development of the nervous systems (Kandel et al. 2000) and have been reproduced in both in vitro (Valentini et al. 1992; Makohliso et al. 1991; Schmidt et al. 1997) and in vivo (Fine et al. 1991; Valentini et al. 1989; Al-Majed et al. 2000) artificial settings. A direct inference is that electrical stimulation can be used clinically to promote nerve repair where nerve regeneration over a substantial distance is required (Sisken et al. 1993; Gordon et al. 2009; Willand et al. 2016). Indeed, research efforts in this direction have been on-going over more than three decades. Depending on the type of electric fields used, there are two classes of electrical stimulations: (1) imposing a directed (constant) DC electric field of a low strength and (2) using AC electrical pulses to trigger APs in the injured neurons. While the exact biophysical mechanisms for these two situations are still a topic of continuing investigation and many hypotheses exist, I think the two approaches work in distinct mechanisms. Weak DC electrical stimulation is more likely to promote neural growth by enhancing the neuron's adhesion to the substrate, which then rides on the mechanism of adhesion traction force-induced cellular growth (Franze and Guck 2010), as changes in cell adhesion can influence multiple cellular pathways including proliferation, differentiation, and migration. In contrast, AC AP stimulation is more likely to trigger Ca^{2+} release in the soma or growing neurite as a second messenger to induce the growth or repair mechanism (Al-Majed et al. 2000; Stewart et al. 1995; Kimura et al. 1998; Ming et al. 2001). Deducing the exact biological mechanisms is beyond the consideration of this chapter and is still a subject of intensive on-going research. The AC AP stimulation is mostly applied to nerves at the tissue level (see Sect. 14.4 for an example application) and is beyond the cellular scope of this book. Therefore, in this chapter, we focus on analyzing the immediate physical effects that a DC electric field imposes on neurons, using equivalent electrical circuit models. This includes how a DC electric field affects neurite outgrowth and how it directs the neurite growth.

L. Guo, *Principles of Electrical Neural Interfacing*,
https://doi.org/10.1007/978-3-030-77677-0_13

13.1   Neuronal Model for Substrate Interaction

A neuron is enclosed by a cell membrane comprising a dielectric phospholipid bilayer, which separates the inner cytoplasma from the outer extracellular space. At rest, the inner membrane is negatively charged at tens of millivolts relative to the outer membrane. The electrical properties of the passive membrane are modeled as a capacitor C_m in parallel with a resistor R_m (Fig. 2.2). R_m is usually hundreds of mega Ohms, and thus the passive neuronal membrane can be electrically approximated as the capacitor C_m alone. When considering a spherical neuron suspended in a solution, its outer surface is positively charged and surrounded by freely movable cations.

To consider the interactions between a neuron and a culture substrate, the cell deposition onto the solid surface can be divided into two stages (Dan 2003): (1) initial physical interactions dominated by long-range, physical forces between the cell and the substrate before cell docking and (2) biophysical interactions taking place at an intimate distance after docking and requiring a longer period of time to stabilize and reinforce. To study these two stages, simply modeling the neuron as a colloidal particle characterized by a positive surface charge, however, is inadequate to account for many of the observed phenomena. This is because the cell surface is enriched with charged glycoproteins that extend tens of nanometers beyond the lipid outer surface. Among such a forest of glycoproteins, small ions are freely penetrable and movable. As a result, when a neuron attaches to a surface, repulsive interactions between these membrane glycoproteins and the ECM proteins, such as fibronectin, adsorbed on the substrate create a cleft of no smaller than 40 nm and often in a range close to 100 nm (Fromherz 2002). For comparison, the Debye length of the EDL (i.e., its thickness) of an electrode is only a few nanometers in aqueous solutions. To account for these membrane characteristics, based on the model proposed by Dan

Fig. 13.1 Core-hair structural model (not to scale) of a neuron to account for the interactions between a neuron and a culture substrate

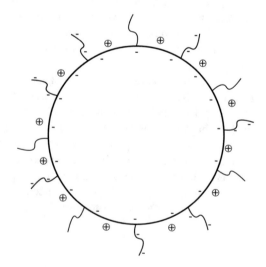

(Dan 2003), as depicted in Fig. 13.1, the neuron is modeled as a *core-hair* structure, where the core represents the soma and the hair layer represents the macromolecular protrusions, including glycoproteins, from the cell membrane. The hair forest is penetrable to small ions in the surrounding medium. Some of the hairs are charged, presumably with a net negative charge. This charged hair layer plays a significant role in the cell's interactions with other cells or a substrate. It is noted that interposed free ions in the charged hair forest reduce the strength of direct electrostatic interactions between the cell and substrate (Dan 2003). Based on this core-hair model, we will qualitatively study the relationships between substrate charges and the corresponding type and strength of neuronal reactions.

13.2 Weak DC Electric Field to Promote Neuronal Growth

It has been shown that neurite outgrowth is significantly enhanced both in vitro and in vivo on positively charged substrates, such as polycationic polymers (e.g., poly-d-lysine (Harnett et al. 2007) and poly-l-lysine (Wu et al. 2020)), electrets[1] (e.g., poled polytetrafluoroethylene (Valentini et al. 1989) and fluorinated ethylene propylene (Makohliso et al. 1991)) and piezoelectric polymers (e.g., poled vinylidenefluoride-trifluoroethylene copolymer (Fine et al. 1991) and polyvinylidene fluoride (Valentini et al. 1992)). In contrast, in a well-controlled *in vitro* study (Wong et al. 1994), the negatively charged (Schmidt et al. 1997) pristine PPy/*p*-toluenesulfonate (PPy/*p*TS) surface could not support attachment and spreading of bovine aortic endothelial cells in serum-free medium, and when the fibronectin-coated PPy/*p*TS substrate was electrically reduced to be more negative for a few hours, attached and spread cells rounded up without change in viability, though in some in vivo cases negatively charged surfaces seemed also to moderately enhance neurite outgrowth due to complicated situations involving modification by positively charged ECM proteins and reconfiguration of the electric fields (Fine et al. 1991; Valentini et al. 1989). Therefore, based on these knowledges, it is reasonable to assume that some of a neuron's extracellular hairs, including the extracellular domain of the ECM anchor proteins—integrins[2], carry a net negative charge, and the neuron can favorably dock, attach, and spread on a positively charged substrate in culture, whereas a negatively charged surface repels the cell from attachment and spreading. When the substrate bears a net negative surface charge, e.g., PPy/pTS (Wong et al. 1994), surface charge modification to the substrate by either adsorbing positively charged ECM proteins

---

[1]Electrets are broadly defined as materials possessing quasi-permanent surface charges because of trapped monopolar charge carriers.

[2]Integrins are a family of heterodimer (alpha and beta subunits) transmembrane proteins that have bi-directional signaling capabilities via their extracellular flexible head domain and the cytoplasmic scaffold proteins.

such as fibronectin[3] (Harnett et al. 2007) or applying a positive potential (Rajnicek et al. 1998) is required to facilitate cell attachment and spreading. For the specific purpose of enhancing neuronal growth, a positive voltage is often used in combination with ECM protein coating to the substrate.

An electric field can be applied during any of the three stages: (1) during ECM protein coating to enhance the adsorption, (2) during cell docking to help attract and stabilize the cell to the substrate, and (3) after cell docking to strengthen the cell-substrate adhesion to promote neurite outgrowth. Taking the neurally favorable PPy material as an example, as PPy's surface is negatively charged (Schmidt et al. 1997), it adsorbed the positively charged ECM protein, fibronectin, more efficiently comparing to neutral tissue-culture polystyrene (TCPS) control; and an imposed electric field to polarize the material during the process further enhanced the adsorption (Kotwal and Schmidt 2001). Using materials carrying permanent positive surface charges may facilitate both ECM protein adsorption and neuronal docking, as well as promoting neurite outgrowth at a later stage, in in vitro (Makohliso et al. 1991; Wu et al. 2020; Valentini et al. 1992) and in vivo (Valentini et al. 1989; Fine et al. 1991) settings. However, in many studies where a more flexible external electrical stimulation approach is used, the DC electrical stimulation is only applied to the substrate or electrode to create a positive voltage after the neurons' attachment (Schmidt et al. 1997). So, we analyze such a situation in more details below. Fig. 13.2 illustrates two in vitro scenarios where a DC electric field is applied to a PPy substrate to enhance the attachment and growth of neurons.

13.2.1   DC Voltage Stimulation

In Fig. 13.2a, a 100 mV positive voltage is applied for two hours between the PPy substrate and a quasi-reference silver (Ag) wire electrode; and a gold (Au) wire counter electrode is placed at a distance away (Schmidt et al. 1997). PC-12 cells (a rat neuronal cell line) have been cultured on the PPy substrate for 24 h before the onset of the electrical stimulation. Although fibronectin is not specifically pre-coated on the PPy substrate, a serum containing medium is used for the culture, so that positive ECM proteins including fibronectin should have been adsorbed on the PPy during the initial 24-h culture to facilitate cell attachment and spreading. Because the applied voltage of 100 mV is within the water electrolysis window (i.e., -0.6 to 0.8 V, see Chap. 3) on Pt surface and PPy is known to generate a lower overpotential η in its EDL than Pt when the same overall potential is applied, no Faradaic current is generated in the EDL, so that the PPy–electrolyte interface functions solely as a capacitor C_{dl} and is in an open-circuit state under the 100 mV DC voltage, which is entirely imposed across the C_{dl}. This 100 mV positive voltage generates additional positive charges in the PPy's backbone. For compensation, additional anions are

---

[3]Fibronectin-coated substrates are referred as monopolar acidic or positively charged.

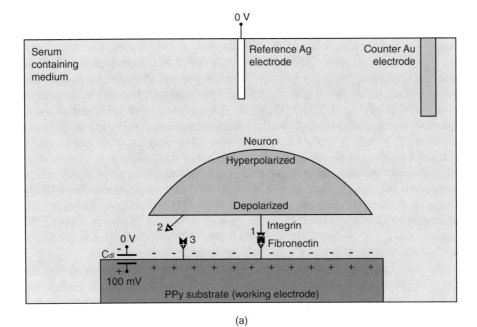

(a)

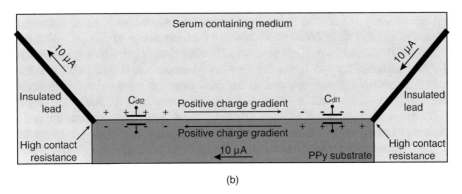

(b)

Fig. 13.2   A DC electric field is applied to a PPy substrate to enhance the attachment and growth of neurons. (**a**) A 100 mV positive voltage is applied between the PPy substrate and a quasi-reference silver (Ag) wire electrode; and a gold (Au) wire counter electrode was placed at a distance away. (**b**) A DC current of 10μA is applied across a PPy film whose resistance is 1 kΩ. Charges in the PPy backbone are shown, whereas the doping ions in the PPy matrix are not shown

electrostatically attracted into the EDL and the cleft between the PPy surface and the cell membrane. Two electrostatic actions are worth our particular attention: (1) the negatively charged extracellular integrin domain (marked as "1" in Fig. 13.2a) which already binds to an adsorbed fibronectin ligand is electrostatically pulled towards the PPy substrate, increasing the traction force to the cytoskeleton; and (2) some unbound integrin terminals (marked as "2") in the hair forest are also electrostatically attracted towards the PPy substrate, where they would have a higher probability to bind the adsorbed available fibronectins (marked as "3") to form new attachments for further spreading of the cell, and thus the net traction force between the neuron and the PPy substrate increases. Over the two-hour period of the imposed electric field, this sustained increase in the cell's net traction force to the substrate presumably activates the genetic pathway for significant enhancement of neurite outgrowth, as observed (Schmidt et al. 1997). Under similar analyses, other evidences also corroborate this hypothesis on adhesion strength-modulated cell growth. When a positive DC voltage is applied to a Pt substrate, protein synthesis in tumor cells is enhanced (Kojima et al. 1992). However, the use of conducting polymers, such as PPy, as the stimulating substrate or electrode is advantageous for a higher charge density comparing to the use of other substrate such as Pt or indium tin oxide (ITO). For a conducting polymer electrode and a Pt or ITO electrode of the same size, according to $Q = CV$, under the same applied voltage V, the much higher EDL capacitance C_{dl} of the conducting polymer electrode gives rise to a much higher surface charge Q that creates a much stronger electric field.

Another hypothesis concerning the enhanced growth mechanism is that the imposed electric field directly depolarizes the cell over the two-hour period, and this sustained depolarization activates the genetic pathway for significant enhancement of cell growth. However, our following analysis rules out this possibility. It is noticed that, in Fig. 13.2a, the lower neuronal membrane is depolarized, as additional anions accumulated in the cleft, whereas the upper membrane is hyperpolarized—the neuron is polarized under the DC voltage from the substrate. In another parallel study (Wong et al. 1994), it is observed that when a DC voltage of -250 mV (which still would not produce a Faradaic current in the EDL) is applied to the PPy substrate for 4–5 h, the cultured cells become round and their growth is halted. In addition to the repulsion to the cells from the more negatively charged substrate, the lower cell membrane is now hyperpolarized, while the upper depolarized. Although half of the cell's membrane is still depolarized, the cell's growth is actually inhibited due to poor substrate adhesion. This evidence rejects this current hypothesis and favors the former hypothesis on adhesion strength-modulated cell growth.

It should also be noted that when the DC voltage is applied between the substrate and the counter electrode in a two-electrode electrochemical cell, i.e., not between the substrate and the reference electrode in a three-electrode electrochemical cell, the voltage is actually split in cascade between the two EDLs of the electrodes. Because in many of the above studies (Wong et al. 1994; Kojima et al. 1992; Schmidt et al. 1997) the smooth metal counter electrode has a much smaller geometry and also a much smaller EDL capacitance than the PPy, Pt, or ITO substrate, the actual voltage allocated across the EDL of the substrate is only a small fraction of the overall voltage (see Exercise 13.2(a)). This could be an explanation to the case that when a -250 mV voltage was applied to the ITO control substrate, cell rounding was not observed (Wong et al. 1994). The actual voltage on the ITO substrate could be too small to exert any observable repulsive effect.

13.2.2 DC Current Stimulation

In the second configuration in Fig. 13.2b, a DC current of 10μA is applied across a PPy film whose resistance is 1 kΩ (Schmidt et al. 1997). An electric field is established within the film between the two sides where electrical leads make contacts. This 10μA current gives rise to a voltage of 10 mV across the PPy film, which is too low to initiate a Faradaic current through the EDL of the PPy–electrolyte interface; and this interface functions electrically as a capacitor, which is in an open-circuit state under the applied DC voltage. Thus, the solution interface actually acts as an insulation to the immersed PPy film.

However, the PPy film is polarized by the applied DC voltage of 10 mV between the two sides. On the anode side where the potential is higher (right side in Fig. 13.2b), anions accumulate both inside the film (not shown in Fig. 13.2b) and in the surrounding solution; and on the cathode side where the potential is lower (left side), cations accumulate both inside the film (not shown) and in the surrounding solution. Based on our analyses in Sect. 13.2.1, neuronal adhesion and growth are enhanced around the anode but inhibited around the cathode. As the applied voltage of 10 mV, and thus the electric field, is very low, the enhancement or inhibition is not substantial. However, the reported enhancement to neurite outgrowth is still significant comparing to the non-stimulated control, only slightly lower than the configuration in Fig. 13.2a (Schmidt et al. 1997). This is presumably due to the following artifact: because the contact resistances between the electrical lead and the PPy film are very high (the leads are attached to the PPy film using electrically conducting, double-sided carbon tape (Kotwal and Schmidt 2001)), given the 10μA DC current, the voltage drop across each of the two contacts is significant, creating a much stronger localized electric field around the anode and cathode, respectively.

13.3 DC Electric Field to Direct Axonal Growth

There are a variety of physical, chemical, and biological cues that can affect the orientation of neurite growth, which has a clear implication in clinical nerve repair. All of these cues are being considered in engineering biomimetic nerve conduits for repairing segmental nerve defects (Daly et al. 2012; De Ruiter et al. 2009). Here, we consider how a directed DC electric field orients and guides the growth of neurites. While several biomolecular mechanisms may be recruited by the electric field and thus responsible for the directed growth (Stewart et al. 1995; Patel and Poo 1982; Rajnicek et al. 1998; Erskine and McCaig 1995), we focus our analysis on how the electric field can directly affect the activity of integrins which are necessary for the formation and appropriate targeting of the growth cone to the directed path.

Under a directed DC electric field through the overlying solution (note, the configuration is different from that in Sect. 13.2.2), not only does the neurite outgrowth rate increase on the cathode side, but the neurites also preferentially

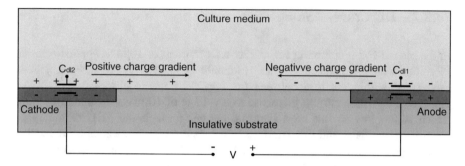

Fig. 13.3   A DC voltage is applied between two substrate-integrated planar electrodes in culture

grow toward the cathode (Patel and Poo 1982; Stewart et al. 1995). In Fig. 13.3, we illustrate a possible direct mechanism to explain these phenomena. A DC electric field is established between two substrate-integrated planar electrodes. If the overpotential on each electrode is within the water electrolysis window of -0.6 to 0.8 V, there should be no DC current flowing across. However, with a sufficient voltage applied, a DC electrical current can be initiated. In this case, the electrodes could be corroded and water around the electrodes could be split. Either without or with the DC current, cations accumulate/adsorb on the substrate around the cathode, which has a lower potential, and anions accumulate/adsorb on the substrate around the anode. Furthermore, a positive surface charge gradient is formed on the substrate, with charge density increasing towards the cathode. Similarly, there is a complementary negative surface charge gradient on the substrate, with charge density increasing towards the anode.

When the substrate itself is positively charged or pre-coated by a positively charged ECM protein (e.g., fibronectin) or polycation (e.g., polylysine), the negatively charged extracellular domain of integrin now attaches better onto the substrate on the cathode side than the anode side, creating a differential force pulling the growth cone toward the cathode (Franze and Guck 2010), and the neurite growth cones follow up the positive surface charge gradient towards the cathode. In the meanwhile, a smaller portion of neurites can also grow on the anode side, as the substrate itself bears positive charges and the anode attracts the negatively charged integrins. In contrast, when the substrate is negatively charged or pre-coated by a negatively charged ECM protein (e.g., laminin), in absence of the externally applied electric field, neurons would attach poorly and remain round on the substrate (Wong et al. 1994). When a DC electric field is applied, cations accumulate and adsorb on the substrate surface around the cathode, with a positive charge gradient pointing from the cathode to the anode, while anions accumulate and adsorb on the substrate around the anode. Thus, with sufficient time, neurons attach and spread near the cathode, and their neurites grow towards the cathode. In contrast, it would be even harder for neurons to attach to and spread on the substrate around the anode. These phenomena are what are observed in a comprehensive study (Rajnicek et al. 1998).

13.4   Summary

- Depending on the type of electric fields used, there are two classes of electrical stimulations: (1) imposing a directed (constant) DC electric field of a low strength and (2) using AC electrical pulses to trigger APs in the injured neurons. Weak DC electrical stimulation is more likely to promote neural growth by enhancing the neuron's adhesion to the substrate, which then rides on the mechanism of adhesion traction force-induced cellular growth. AC AP stimulation is more likely to trigger Ca^{2+} release in the soma or growing neurite as a second messenger to induce the growth or repair mechanism.
- To account for the interactions between a neuron and a culture substrate, a core-hair structural model of a neuron is proposed. The core represents the soma and the hair layer represents the macromolecular protrusions, including glycoproteins, from the cell membrane. The hair forest is penetrable to small ions in the surrounding medium. Some of the hairs are charged, presumably with a net negative charge. This charged hair layer plays a significant role in the cell's interactions with other cells or a substrate.
- Some of a neuron's extracellular hairs, including the extracellular domain of the ECM anchor proteins—integrins, carry a net negative charge, and the neuron can favorably dock, attach, and spread on a positively charged substrate in culture, whereas a negatively charge surface repels the cell from attachment and spreading. When the substrate bears a net negative surface charge, surface charge modification to the substrate by either adsorbing positively charged ECM proteins such as fibronectin or applying a positive potential is required to facilitate cell attachment and spreading. For the specific purpose of enhancing neuronal growth, a positive voltage is often used in combination with ECM protein coating to the substrate.
- Under a 100 mV positive DC voltage, the PPy-electrolyte interface functions solely as an EDL capacitor and is in an open-circuit state where the 100 mV voltage is entirely imposed across the EDL capacitor. This voltage generates additional positive charges in the PPy's backbone and additional anions in the EDL and the cleft between the PPy surface and the cell membrane. Through electrostatic interactions, the net traction force between the neuron and the PPy substrate increases, which activates the genetic pathway for significant enhancement of neurite outgrowth. The use of conducting polymers as the stimulating substrate is advantageous for a higher charge density.
- When a DC current of 10µA is applied across a PPy film, the PPy–electrolyte interface functions electrically as a capacitor, which is in an open-circuit state. An electric field is established within the film and polarizes it. Neuronal adhesion and growth are enhanced around the anode but inhibited around the cathode.
- When a DC voltage is applied between two substrate-integrated planar electrodes in culture, cations accumulate/adsorb on the substrate around the cathode; and anions accumulate/adsorb on the substrate around the anode. A positive surface

charge gradient is formed on the substrate, with charge density increasing towards the cathode. Neurons attach preferentially on the cathode side, and the neurite growth cones follow up the positive surface charge gradient towards the cathode.

Exercises

13.1 Please describe the limitations of the core-hair neuronal model for substrate interaction.

13.2 When a material's surface has a net negative charge, why cannot most neurons attach and spread on it? What can you do to culture neurons on such a surface?

13.3 In Fig. 13.2a, if the 100 mV DC voltage is applied between the substrate and the counter Au electrode, the voltage is actually split in cascade between the two EDLs of the electrodes. Please answer the following questions:

(a) Given the EDL capacitances of the working and counter electrodes as C_{dlw} and C_{dlc}, respectively, with $C_{dlw} = 100C_{dlc}$. Derive the DC voltage across the EDL capacitance C_{dlw} of the working electrode. What can you infer from this result then?

(b) Explain why the overlaid neuron has its lower membrane facing the substrate depolarized, while the upper membrane hyperpolarized.

13.4 Comparing Figs. 13.2b and 13.3, explain why neuronal adhesion and growth are enhanced around the anode but inhibited around the cathode in Fig. 13.2b, whereas the opposite holds in Fig. 13.3.

13.5 Refer to the electrical stimulation setup below, where the two stimulating electrodes are suspended in the medium over the culture substrate. Suppose the two electrodes are placed far apart and far away from the neuron and the substrate. Analyze the equivalent electrical circuit and answer the following questions:

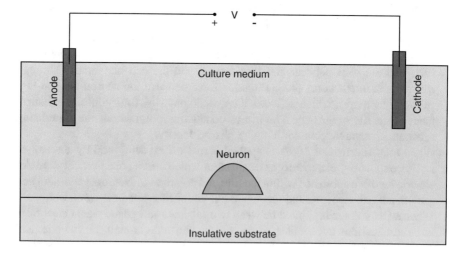

(a) When a DC voltage of 100 mV is applied as shown, what direct effect is caused to the neuron?
(b) When the stimulus is changed to a 10μA DC current, what is the result then?

13.6 Based on the knowledge learned in this chapter, how do you think weak DC electrical stimulation can be applied in vivo to promote nerve regeneration?

References

Al-Majed AA, Neumann CM, Brushart TM, Gordon T (2000) Brief electrical stimulation promotes the speed and accuracy of motor axonal regeneration. Journal of Neuroscience 20 (7):2602-2608

Daly W, Yao L, Zeugolis D, Windebank A, Pandit A (2012) A biomaterials approach to peripheral nerve regeneration: bridging the peripheral nerve gap and enhancing functional recovery. Journal of the Royal Society Interface 9 (67):202-221

Dan N (2003) The effect of charge regulation on cell adhesion to substrates: salt-induced repulsion. Colloids and Surfaces B: Biointerfaces 27 (1):41-47

De Ruiter GC, Malessy MJ, Yaszemski MJ, Windebank AJ, Spinner RJ (2009) Designing ideal conduits for peripheral nerve repair. Neurosurgical focus 26 (2):E5

Erskine L, McCaig CD (1995) Growth cone neurotransmitter receptor activation modulates electric field-guided nerve growth. Dev Biol 171 (2):330-339

Fine EG, Valentini RF, Bellamkonda R, Aebischer P (1991) Improved nerve regeneration through piezoelectric vinylidenefluoride-trifluoroethylene copolymer guidance channels. Biomaterials 12 (8):775-780

Franze K, Guck J (2010) The biophysics of neuronal growth. Reports on Progress in Physics 73 (9):094601

Fromherz P (2002) Electrical interfacing of nerve cells and semiconductor chips. Chemphyschem 3 (3):276-284. doi:https://doi.org/10.1002/1439-7641(20020315)

Gordon T, Sulaiman OA, Ladak A (2009) Electrical stimulation for improving nerve regeneration: where do we stand? International review of neurobiology 87:433-444

Harnett EM, Alderman J, Wood T (2007) The surface energy of various biomaterials coated with adhesion molecules used in cell culture. Colloids and surfaces B: Biointerfaces 55 (1):90-97

Kandel ER, Schwartz JH, Jessell TM, Biochemistry Do, Jessell MBT, Siegelbaum S, Hudspeth A (2000) Principles of neural science, vol 4. McGraw-Hill, New York,

Kimura K, Yanagida Y, Haruyama T, Kobatake E, Aizawa M (1998) Electrically induced neurite outgrowth of PC12 cells on the electrode surface. Medical and Biological Engineering and Computing 36 (4):493-498

Kojima J, Shinohara H, Ikariyama Y, Aizawa M, Nagaike K, Morioka S (1992) Electrically promoted protein production by mammalian cells cultured on the electrode surface. Biotechnol Bioeng 39 (1):27-32. doi:https://doi.org/10.1002/bit.260390106

Kotwal A, Schmidt CE (2001) Electrical stimulation alters protein adsorption and nerve cell interactions with electrically conducting biomaterials. Biomaterials 22 (10):1055-1064. doi:https://doi.org/10.1016/s0142-9612(00)00344-6

Makohliso S, Valentini R, West J, Aebischer P Positive electret substrates enhance neuroblastoma process outgrowth. In: [1991 Proceedings] 7th International Symposium on Electrets (ISE 7), 1991. IEEE, pp 712-716

Ming G, Henley J, Tessier-Lavigne M, Song H, Poo M (2001) Electrical activity modulates growth cone guidance by diffusible factors. Neuron 29 (2):441-452. doi:https://doi.org/10.1016/s0896-6273(01)00217-3

Patel N, Poo M-M (1982) Orientation of neurite growth by extracellular electric fields. Journal of Neuroscience 2 (4):483-496

Rajnicek AM, Robinson KR, McCaig CD (1998) The direction of neurite growth in a weak DC electric field depends on the substratum: contributions of adhesivity and net surface charge. Dev Biol 203 (2):412-423

Schmidt CE, Shastri VR, Vacanti JP, Langer R (1997) Stimulation of neurite outgrowth using an electrically conducting polymer. Proceedings of the National Academy of Sciences 94 (17):8948-8953

Sisken BF, Walker J, Orgel M (1993) Prospects on clinical applications of electrical stimulation for nerve regeneration. Journal of cellular biochemistry 51 (4):404-409

Stewart R, Erskine L, McCaig CD (1995) Calcium channel subtypes and intracellular calcium stores modulate electric field-stimulated and-oriented nerve growth. Dev Biol 171 (2):340-351

Valentini R, Sabatini A, Dario P, Aebischer P (1989) Polymer electret guidance channels enhance peripheral nerve regeneration in mice. Brain Res 480 (1-2):300-304

Valentini RF, Vargo TG, Gardella Jr JA, Aebischer P (1992) Electrically charged polymeric substrates enhance nerve fibre outgrowth in vitro. Biomaterials 13 (3):183-190

Willand MP, Nguyen M-A, Borschel GH, Gordon T (2016) Electrical stimulation to promote peripheral nerve regeneration. Neurorehabilitation and Neural Repair 30 (5):490-496

Wong JY, Langer R, Ingber DE (1994) Electrically conducting polymers can noninvasively control the shape and growth of mammalian cells. Proceedings of the National Academy of Sciences 91 (8):3201-3204

Wu Y, Wang M, Wang Y, Yang H, Qi H, Seicol BJ, Xie R, Guo L (2020) A neuronal wiring platform through microridges for rationally engineered neural circuits. APL Bioengineering 4 (4):046106

Part IV
Applications

Chapter 14
Applications

To a lay person, the most intriguing aspect of a technical field would be the type of exemplar applications. And indeed, many people first learn this field of neuroengineering actually from its applications, like BCIs. From a developmental perspective, the initial cultivation of this field was also driven by prominent practical applications such as BCIs and functional electrical stimulation (FES). Over the years, many young people have been attracted to this new field by these intriguing applications. While a number of applications have ultimately matured as commercial products, up to now, the majority are still in the early research phase, and still others are in the visionary phase as frequently seen in science-fiction movies.

While most knowledge we have covered in this book are directly applicable in neuroelectrophysiological research, in this chapter, we go beyond to describe four exemplar applications, two for neural recording and two for neural stimulation, respectively. For each neural interfacing mode, we also include a present industrial/clinical application. However, the true excitement of this rising field lies in that, in the next few decades, many promising applications are likely to advance from the research phase to become commercial products that can benefit our daily lives.

14.1 High-Performance BCIs

BCIs represent the most intriguing neuroengineering application ever since the beginning of this field. Actually, the pursuit of BCIs is one of the major factors contributing to the emergence of this field. The concept of BCI was first proposed by Dr. Jacques J. Vidal in 1976 (Vidal 1977). Since then, non-invasive electroencephalogram (EEG) based BCIs had dominated the subfield until the year of 2000, for their easier acceptability by human subjects. And up to now, EEG-based BCIs are still an active area of study. A number of consumer BCI devices based on EEG had also emerged in late 2000. However, EEG-based BCIs have limited performances with regard to speed, accuracy, and complexity, due to the limited bandwidth of

© The Author(s), under exclusive license to Springer Nature Switzerland AG 2022
L. Guo, *Principles of Electrical Neural Interfacing*,
https://doi.org/10.1007/978-3-030-77677-0_14

EEG. Invasive BCIs based on intracortical single-unit recording, though technically more challenging and stringent, represent the highest level of performances that BCIs can achieve. Thus, here we focus on this subclass of BCIs.

In 1978, the first invasive BCI for a vision prosthesis targeting the visual cortex was built to aid the blind (Dobelle et al. 1979). This was a stimulatory prosthesis that mapped video images to electrical stimulation patterns and delivered them onto the visual cortex using a surface platinum electrode array. Over the next few decades, the field has concentrated on motor prostheses, for the apparent motivation to restore basic motor functions to quadriplegic patients. In the turning of 2000, intracortical recording based BCIs were first implemented in humans using a hand-crafted polytrode (Bartels et al. 2008), allowing the subject to drive a cursor on a computer monitor for typing and speech synthesis (Kennedy et al. 2000). Then, multiunit recording based BCIs exploded in the following two decades, primarily enabled by the micromachined high-density UEA (Fig. 14.1a, b). Initially, it was demonstrated that multichannel neural activities recorded from the motor cortex of a monkey during free arm movements could be used to replicate the movements on a robotic arm in real time, both locally and through the internet (Wessberg et al. 2000). Later, it was shown that a monkey could learn to use its motor cortical activities to control a prosthetic arm to feed itself (Velliste et al. 2008). In the meantime, a tetraplegic human subject was able to perform rudimentary actions including controlling a cursor on a computer screen and a multi-jointed robotic arm, simply using his natural movement intents (Fig. 14.1) (Hochberg et al. 2006). This BCI recorded natural movement intents in the primary motor cortex using the UEA. Earlier in 2002, a startup company called Cyberkinetics was founded with the ambition to commercialize this exciting technology. The sophistication of controlling a robotic arm by a tetraplegic patient was further improved a few years later (see Fig. 14.2) (Hochberg et al. 2012; Collinger et al. 2013). All these efforts ultimately pushed the UEA to be cleared for clinical use by FDA; and up to now, it remains the only penetrating MEA in the world that is FDA cleared. More recent studies extended the use of natural movement intents to continuously control the quadriplegic patient's own paralyzed arm and hand in real time through FES, restoring both reaching and grasping movements (see Fig. 14.3a, b) (Bouton et al. 2016; Ajiboye et al. 2017), while the latest study restored the sense of touch to the quadriplegic patient during reanimated motor functions (Fig. 14.3c) (Ganzer et al. 2020). Resonating with these exciting scientific advances, a new wave of entrepreneurship has also kicked into this promising technological venture with startups including Neuralink and Paradromics.

14.2 Drug Screening

Planar MEAs integrated in cell-culture wells (see Fig. 14.4 for an example) have been established over the past two decades as a robust tool for standard in vitro neurotoxicity/neuropharmacology evaluations to screen selected compounds under physiological or induced pathophysiological conditions in the pharmaceutical

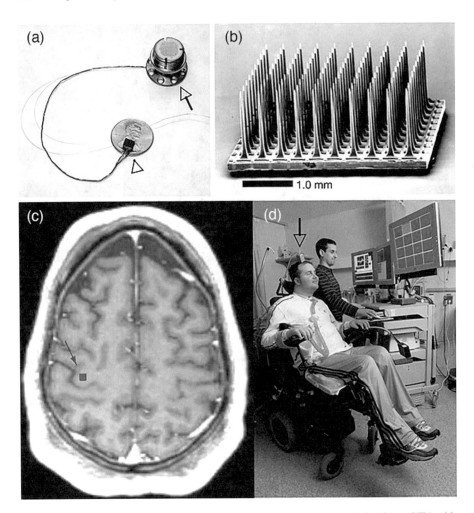

Fig. 14.1 First reported human BCI using the UEA. (**a**) Overview of the brain implant: a UEA with connecting wires and a skull-mounting connector. (**b**) A close view of the UEA. (**c**) MRI image showing the intended implantation site for the UEA (red arrow). (**d**) A tetraplegic patient with the BCI controlling a cursor on a computer screen using his movement intents. Reproduced with permission from (Hochberg et al. 2006). Copyright © 2006 Nature Publishing Group. Courtesy of www.braingate.org

industry (Novellino et al. 2011; Shafer 2019; Morefield et al. 2000). A variety of specimens, including primary neurons, cardiac cells, muscle cells, different cell lines, stem cells, different tissue slices, organoids, and even specifically engineered organs on a chip from both animals and humans sources, can be cultured on top of the MEA for extracellular monitoring over a long period of time. Multichannel recordings spanning an extended area can be obtained to evaluate the spatiotemporal activity patterns across the specimen. Simultaneous electrical stimulation can also be employed to study perturbation-induced activity patterns. In neurotoxicity testing,

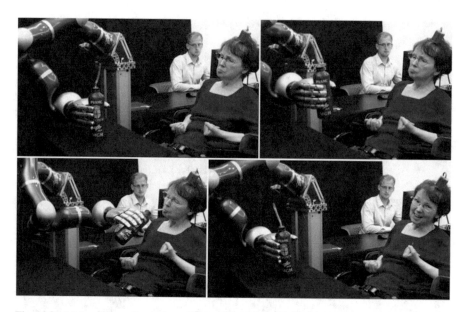

Fig. 14.2 A tetraplegic patient using a BCI to control a sophisticated robotic arm and hand for self-drinking, simply by her natural movement intents. A UEA was implanted in the arm area of her motor cortex to record the neural activities associated with her movement intents. Reproduced with permission from (Hochberg et al. 2012). Copyright © 2012 Macmillan Publishers Limited. Courtesy of www.braingate.org

both spike waveforms and burst patterns of the cultured neuronal network can be affected by the treatment with specific neuroactive chemicals in the culture medium; and the induced changes are usually detected by the underlying MEA. To increase the SNR of the recordings and detect subthreshold activities, 3D microelectrode profiles were developed to tighten the interfacial seal between the electrode and cell membrane (Didier et al. 2020).

14.3   Chronic Pain Management

Electrical stimulation has been widely explored as a reversible therapy for the treatment of various chronic pain syndromes. This includes using non-invasive brain stimulation (O'Connell et al. 2018), deep brain stimulation (Boccard et al. 2015; Rasche et al. 2006), spinal cord stimulation (Cameron 2004; de Leon-Casasola 2009), transcutaneous nerve stimulation (Mokhtari et al. 2020), and peripheral nerve stimulation using an implanted stimulator (Fig. 14.5) (de Leon-Casasola 2009; Reverberi et al. 2014; Slavin 2008; Mobbs et al. 2007). Specifically, for peripheral nerve stimulation, kilohertz frequency alternating current (KHFAC) applied via a bipolar cuff electrode around the nerve is used to interrupt the AP conduction along the tonically misfiring pain fibers. The conduction block has been verified not due to

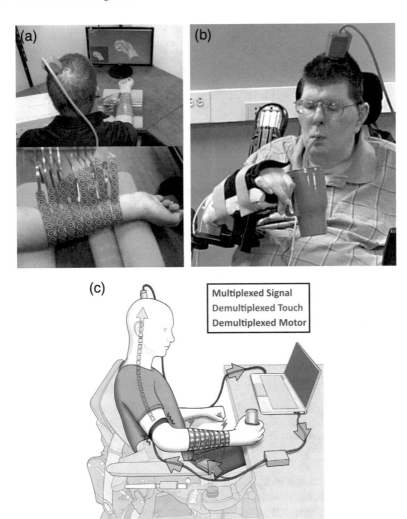

Fig. 14.3 Real-time BCI control of paralyzed arm and hand. (**a**) A quadriplegic patient controlled his own paralyzed arm and hand in real time through BCI+FES. Reproduced with permission from (Bouton et al. 2016). Copyright © 2016 Macmillan Publishers Limited. (**b**) Another quadriplegic patient controlled his own paralyzed arm and hand through BCI+FES to drink a cup of coffee. Reproduced with permission from (Ajiboye et al. 2017). Copyright © 2017 Elsevier Inc. Courtesy of Case Western Reserve University. (**c**) Illustration showing the sense of touch was restored to a quadriplegic patient during his reanimated motor functions. Reproduced with permission from (Ganzer et al. 2020). Copyright © 2020 Elsevier Inc

effector fatigue or neurotransmitter depletion but is a true block of the AP generation around the stimulating electrodes. The conduction block can be immediately reversed without affecting the nerve, and long-term stimulation using biphasic stimuli appears not to cause discernable damage to the nerve, either. Thus, this approach represents a promising therapeutic intervention in treating chronic

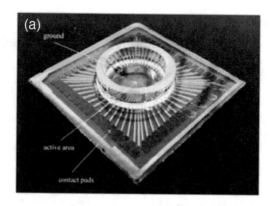

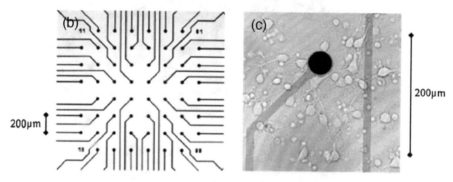

Fig. 14.4 A planar MEA integrated in a cell-culture well. (**a**) Overview of the MEA. (**b**) Microscopic view of the center of the MEA, showing the 60 microelectrodes. (**c**) A closeup view of a planar microelectrode surrounded by dissociated neurons in culture. Reproduced with permission from (Warwick 2014). Copyright © 2014 Springer Science+Business Media Dordrecht

neuropathic pain (Colloca et al. 2017; Slavin 2008; Mobbs et al. 2007). Ongoing research is still focusing on elucidating the biophysical mechanisms underlying the conduction block, which will inform the optimal designs of stimulation parameters and electrode configurations for efficacious clinical applications.

14.4 Nerve Regeneration

In Chap. 13, we have discussed how a weak constant DC electric field enhances the neuronal growth and directs the axonal growth. In this chapter, we look at how electrical stimulation in general, including both DC and AC electrical stimulations, is being investigated as a potential therapeutic approach in promoting the repair of peripheral nerve segmental defects. Because this approach is still being evaluated in the pre-clinical stage, we focus our discussion on research using animal (mostly rodents) segmental nerve defect models.

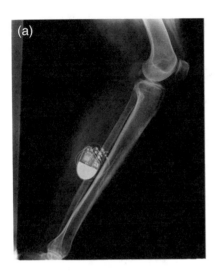
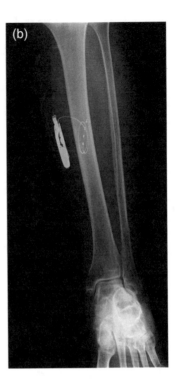

Fig. 14.5 X-ray image of an implanted nerve stimulator in the lower leg of a patient. The stimulator (Lightpulse 100, Neurimpulse, Rubano, PD, Italy) was battery powered. The electrode leads were placed on the tibial nerve. (**a**) Side view. (**b**) Front view. Reproduced with permission from (Reverberi et al. 2014). Copyright © 2014 International Neuromodulation Society

Both DC and AC electrical stimulations have been shown to promote peripheral nerve regeneration in segmental nerve defect animal models. Electrets and piezo-electric polymers as component materials in nerve guidance conduits have been shown to enhance nerve regeneration in rodent models (Valentini et al. 1989; Fine et al. 1991). AC electrical stimulation has also been shown to promote neurite outgrowth in vitro (Ming et al. 2001; Kimura et al. 1998) and enhance nerve regeneration in rodent models (Al-Majed et al. 2000). Here, we describe a latest cutting-edge implantable engineering design to implement this AC approach for short-term in vivo stimulation in a rodent model using an innovative wireless bioresorbable electronic system (Koo et al. 2018; Choi et al. 2020).

It has been observed that even brief AC electrical stimulation applied to the proximal nerve terminal at the beginning of nerve repair can substantially enhance the functional recovery (Al-Majed et al. 2000). And many other studies concur that electrical stimulation applied during the early stage after the nerve repair is sufficient to promote the nerve regeneration, and that the later stage stimulation produces diminishing effects (Gordon 2016). For this type of short-term implanted application, the implanted electronic device needs to be explanted after the nerve

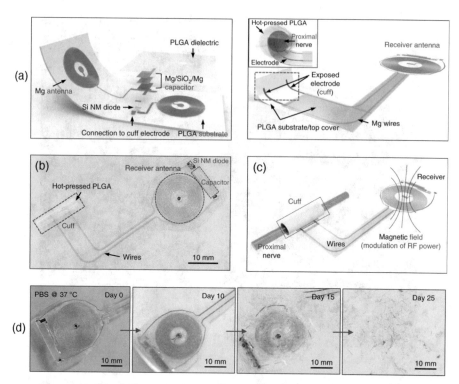

Fig. 14.6 A bioresorbable, wireless electrical nerve stimulator used to apply a short-term electrical stimulation therapy to a repaired nerve after implantation. (**a**) Schematic illustration of the device design: left, an RF power harvester (Folding the constructed system in half yields a compact device with a double-coil inductor); right, the electrode and cuff interface for nerve stimulation (Rolling the end of the system into a cylinder creates a cuff with exposed electrodes at the ends as an interface to the nerve). (**b**) Image of a completed device. (**c**) Schematic of wireless operation, including the nerve interface. (**d**) Images of the dissolution process of a device in PBS (pH = 7.4) at 37 °C. Reproduced with permission from (Koo et al. 2018). Copyright © 2018 Springer Nature

regeneration and reinnervation, which requires a second surgery that causes additional pain and trauma. However, if the implant can dissolve or be resorbed by the body after the initial period of operation, a second surgery would be exempted. This is the concept of *transient medical device*. Figure 14.6 shows such a transient medical device for promoting nerve regeneration—a wireless bioresorbable electronic stimulator (Koo et al. 2018).

Bioresorbable materials were used to fabricate the implantable electronic stimulator, comprising a receiver antenna, a rectifying diode and capacitor, leads, electrodes, substrate, and encapsulation. For implantation, the receiver antenna was placed subcutaneously, while the leads connected to the cuff electrodes on a deeper nerve. Radio frequency (RF) magnetic coupling (~5 MHz) was used to transmit cathodic, monophasic electrical impulses (200μs, 100–300 mV) to excite the target nerve. The highlight of this system was that the constituent materials were bioresorbed in a controlled manner within a defined time frame after subcutaneous implantation. Specifically, constituent materials dissolved within 3 weeks, while all remaining residues completely disappeared after 25 days.

A sciatic nerve transection rat model was used to evaluate the performance of this transient electronic stimulator. The results showed that when assessed at 10 weeks after the injury and repair, repeated 1-h daily electrical stimulation for a total duration of 6 days during the early stage demonstrated enhanced therapeutic effects in terms of the rate and degree of nerve regeneration and recovery of muscle function, comparing to solely applying the electrical stimulation during the repair surgery.

Currently, this cool technology is still under development with regard to adjusting the materials' degradation rates to match the clinically relevant timeframe and optimizing the mechanical and biochemical properties of the constituent materials for in vivo performances (Koo et al. 2018). The first clinical attempt may still need to wait for at least another decade.

Exercises

14.1 Please describe some other applications related to the knowledge we have learned in this book.

14.2 Parkinson's disease is a chronic and progressive movement disorder. Please describe (you may include figures):

(a) The behavioral and cognitive symptoms.
(b) The current understanding on the mechanism of cause from systems, cellular and molecular levels, respectively.
(c) The principle of DBS treatment for drug-resistant Parkinson's disease.
(d) The state-of-the-art DBS technology, including electrodes, stimulators, etc.
(e) What aspects do you think the current DBS technology needs to improve in?

14.3 Search on the internet to find a BCI company, for example Neuralink or Paradromics:

(a) Describe the major intended product of the company, including the target consumers and current status of the product development.
(b) Describe the technological rationales, principles, or mechanisms for the product's functional design.
(c) What are the major technological hurdles for the product to ultimately reach to its intended consumer adoption?
(d) What aspects do you think the current product needs to improve in?
(e) In a broader context, can you identify some of the competitors to the company and articulate the unique competences of its product?

References

Ajiboye AB, Willett FR, Young DR, Memberg WD, Murphy BA, Miller JP, Walter BL, Sweet JA, Hoyen HA, Keith MW (2017) Restoration of reaching and grasping movements through brain-controlled muscle stimulation in a person with tetraplegia: a proof-of-concept demonstration. The Lancet 389 (10081):1821-1830

Al-Majed AA, Neumann CM, Brushart TM, Gordon T (2000) Brief electrical stimulation promotes the speed and accuracy of motor axonal regeneration. Journal of Neuroscience 20 (7):2602-2608

Bartels J, Andreasen D, Ehirim P, Mao H, Seibert S, Wright EJ, Kennedy P (2008) Neurotrophic electrode: method of assembly and implantation into human motor speech cortex. J Neurosci Meth 174 (2):168-176

Boccard SG, Pereira EA, Aziz TZ (2015) Deep brain stimulation for chronic pain. Journal of Clinical Neuroscience 22 (10):1537-1543

Bouton CE, Shaikhouni A, Annetta NV, Bockbrader MA, Friedenberg DA, Nielson DM, Sharma G, Sederberg PB, Glenn BC, Mysiw WJ, Morgan AG, Deogaonkar M, Rezai AR (2016) Restoring cortical control of functional movement in a human with quadriplegia. Nature 533 (7602):247-250. doi:https://doi.org/10.1038/nature17435

Cameron T (2004) Safety and efficacy of spinal cord stimulation for the treatment of chronic pain: a 20-year literature review. Journal of Neurosurgery: Spine 100 (3):254-267

Choi YS, Hsueh YY, Koo J, Yang Q, Avila R, Hu B, Xie Z, Lee G, Ning Z, Liu C, Xu Y, Lee YJ, Zhao W, Fang J, Deng Y, Lee SM, Vazquez-Guardado A, Stepien I, Yan Y, Song JW, Haney C, Oh YS, Liu W, Yun HJ, Banks A, MacEwan MR, Ameer GA, Ray WZ, Huang Y, Xie T, Franz CK, Li S, Rogers JA (2020) Stretchable, dynamic covalent polymers for soft, long-lived bioresorbable electronic stimulators designed to facilitate neuromuscular regeneration. Nat Commun 11 (1):5990. doi:https://doi.org/10.1038/s41467-020-19660-6

Collinger JL, Wodlinger B, Downey JE, Wang W, Tyler-Kabara EC, Weber DJ, McMorland AJ, Velliste M, Boninger ML, Schwartz AB (2013) High-performance neuroprosthetic control by an individual with tetraplegia. The Lancet 381 (9866):557-564

Colloca L, Ludman T, Bouhassira D, Baron R, Dickenson AH, Yarnitsky D, Freeman R, Truini A, Attal N, Finnerup NB (2017) Neuropathic pain. Nature Reviews Disease Primers 3 (1):1-19

de Leon-Casasola OA (2009) Spinal cord and peripheral nerve stimulation techniques for neuropathic pain. Journal of Pain and Symptom Management 38 (2):S28-S38

Didier CM, Kundu A, DeRoo D, Rajaraman S (2020) Development of In Vitro 2D and 3D Microelectrode Arrays and Their Role in Advancing Biomedical Research. Journal of Micromechanics and Microengineering

Dobelle WH, Dobelle WH, Quest DO, Antunes JL, Roberts TS, Girvin JP (1979) Artificial vision for the blind by electrical stimulation of the visual cortex. Neurosurgery 5 (4):521-527

Fine EG, Valentini RF, Bellamkonda R, Aebischer P (1991) Improved nerve regeneration through piezoelectric vinylidenefluoride-trifluoroethylene copolymer guidance channels. Biomaterials 12 (8):775-780

Ganzer PD, Colachis SC 4th, Schwemmer MA, Friedenberg DA, Dunlap CF, Swiftney CE, Jacobowitz AF, Weber DJ, Bockbrader MA, Sharma G (2020) Restoring the Sense of Touch Using a Sensorimotor Demultiplexing Neural Interface. Cell

Gordon T (2016) Electrical stimulation to enhance axon regeneration after peripheral nerve injuries in animal models and humans. Neurotherapeutics 13 (2):295-310

Hochberg LR, Bacher D, Jarosiewicz B, Masse NY, Simeral JD, Vogel J, Haddadin S, Liu J, Cash SS, Van Der Smagt P (2012) Reach and grasp by people with tetraplegia using a neurally controlled robotic arm. Nature 485 (7398):372-375

Hochberg LR, Serruya MD, Friehs GM, Mukand JA, Saleh M, Caplan AH, Branner A, Chen D, Penn RD, Donoghue JP (2006) Neuronal ensemble control of prosthetic devices by a human with tetraplegia. Nature 442 (7099):164-171. doi:https://doi.org/10.1038/nature04970

Kennedy PR, Bakay RA, Moore MM, Adams K, Goldwaithe J (2000) Direct control of a computer from the human central nervous system. IEEE Transactions on Rehabilitation Engineering 8 (2):198-202

Kimura K, Yanagida Y, Haruyama T, Kobatake E, Aizawa M (1998) Electrically induced neurite outgrowth of PC12 cells on the electrode surface. Medical and Biological Engineering and Computing 36 (4):493-498

Koo J, MacEwan MR, Kang SK, Won SM, Stephen M, Gamble P, Xie Z, Yan Y, Chen YY, Shin J, Birenbaum N, Chung S, Kim SB, Khalifeh J, Harburg DV, Bean K, Paskett M, Kim J, Zohny ZS, Lee SM, Zhang R, Luo K, Ji B, Banks A, Lee HM, Huang Y, Ray WZ, Rogers JA (2018) Wireless bioresorbable electronic system enables sustained nonpharmacological neuroregenerative therapy. Nature Medicine 24 (12):1830-1836. doi:https://doi.org/10.1038/s41591-018-0196-2

Ming G, Henley J, Tessier-Lavigne M, Song H, Poo M (2001) Electrical activity modulates growth cone guidance by diffusible factors. Neuron 29 (2):441-452. doi:https://doi.org/10.1016/s0896-6273(01)00217-3

Mobbs R, Nair S, Blum P (2007) Peripheral nerve stimulation for the treatment of chronic pain. Journal of Clinical Neuroscience 14 (3):216-221

Mokhtari T, Ren Q, Li N, Wang F, Bi Y, Hu L (2020) Transcutaneous electrical nerve stimulation in relieving neuropathic pain: Basic mechanisms and clinical applications. Current Pain and Headache Reports 24 (4):1-14

Morefield S, Keefer E, Chapman K, Gross G (2000) Drug evaluations using neuronal networks cultured on microelectrode arrays. Biosensors and Bioelectronics 15 (7-8):383-396

Novellino A, Scelfo B, Palosaari T, Price A, Sobanski T, Shafer TJ, Johnstone AF, Gross GW, Gramowski A, Schroeder O (2011) Development of micro-electrode array based tests for neurotoxicity: assessment of interlaboratory reproducibility with neuroactive chemicals. Frontiers in Neuroengineering 4:4

O'Connell NE, Marston L, Spencer S, DeSouza LH, Wand BM (2018) Non-invasive brain stimulation techniques for chronic pain. Cochrane database of systematic reviews (3)

Rasche D, Rinaldi PC, Young RF, Tronnier VM (2006) Deep brain stimulation for the treatment of various chronic pain syndromes. Neurosurgical focus 21 (6):1-8

Reverberi C, Dario A, Barolat G, Zuccon G (2014) Using peripheral nerve stimulation (PNS) to treat neuropathic pain: a clinical series Neuromodulation: Technology at the Neural Interface 17 (8):777-783

Shafer TJ (2019) Application of microelectrode array approaches to neurotoxicity testing and screening. In: In Vitro Neuronal Networks. Springer, pp 275–297

Slavin KV (2008) Peripheral nerve stimulation for neuropathic pain. Neurotherapeutics 5 (1):100-106

Valentini R, Sabatini A, Dario P, Aebischer P (1989) Polymer electret guidance channels enhance peripheral nerve regeneration in mice. Brain Res 480 (1-2):300-304

Velliste M, Perel S, Spalding MC, Whitford AS, Schwartz AB (2008) Cortical control of a prosthetic arm for self-feeding. Nature 453 (7198):1098-1101. doi:https://doi.org/10.1038/nature06996

Vidal JJ (1977) Real-time detection of brain events in EEG. P IEEE 65 (5):633-641

Warwick K (2014) The cyborg revolution. Nanoethics 8 (3):263-273

Wessberg J, Stambaugh CR, Kralik JD, Beck PD, Laubach M, Chapin JK, Kim J, Biggs SJ, Srinivasan MA, Nicolelis MA (2000) Real-time prediction of hand trajectory by ensembles of cortical neurons in primates. Nature 408 (6810):361-365. doi:https://doi.org/10.1038/35042582

Index

© The Author(s), under exclusive license to Springer Nature Switzerland AG 2022
L. Guo, *Principles of Electrical Neural Interfacing*,
https://doi.org/10.1007/978-3-030-77677-0

Printed in the United States
by Baker & Taylor Publisher Services